COMPLETE GUIDE TO
RUTAN HOMEBUILT AIRCRAFT

by Don and Julia Downie

MODERN AVIATION SERIES

TAB BOOKS Inc.

BLUE RIDGE SUMMIT, PA. 17214

To the hard-working homebuilders reaching for affordable flying through innovative concepts; may they continue to build the airframes with new designs, leading the way to aviation's future.

FIRST EDITION

SECOND PRINTING

Printed in the United States of America

Library of Congress Cataloging in Publication Data

Downie, Don.
    Complete guide to Rutan homebuilt aircraft.

    Includes index.
    1. Airplanes, Home-built. I. Downie, Julia, joint author. II. Title.
TL671.2.D63   629.133'340422   80-28778
ISBN 0-8306-2310-8 (pbk.)

Cover photo Mike Melvill in VariViggen over the Colorado River near Bullhead City, Arizona.

Contents

Introduction

We first met Burt Rutan at a mini-symposium of the Society of Experimental Test Pilots in San Diego, California, where both of us were on a speaker's panel. The next time we saw him was at Brackett Field near Los Angeles, California, where Burt was just beginning to demonstrate his new twin-engine Defiant to the aviation press. At that time, the eminently successful "push-pull" had only 30 hours flight time and was barely permitted into the Los Angeles area. Our flight report, carried in *AOPA Pilot* Magazine, was a "scoop" by an edition or two over other aviation publications.

We approached Burt about doing this book on several occasions before he found enough time in his schedule to give us the green light. At that time, he supplied copies of his voluminous quarterly builders publication, *"Canard Pusher,"* a detailed package of how-to-do-it instructions he had prepared to educate new builders in the fine art of foam and fiberglass structures. We waded through a great many magazine articles—some accurate and some very "blue sky"—about Rutan who is a designer unique in homebuilt aircraft circles.

We flew with Burt in his other designs. We flew with his brother Dick in the popular Long-EZ prototype and then went ahead to fly the ship solo. We flew in the No. 2 VariViggen with builder Mike Melvill. We took Rutan's parents, George and Irene, in the back of our Cessna 170B to a VariEze Fly-in at Bullhead, Arizona, and absorbed the flavor and enthusiasm of the growing family of builders who share both flying and fabrication experiences with each other. We "kicked tires" and admired the meticulous detail of many homebuilders who have vested unending hours in work and rework to come out with a product of which they are truly proud—as they should be.

Certainly one of the more enjoyable parts of preparation of this material has been the many flights up over the Sierra Madre Mountains to Mojave, California, where Rutan's RAF (Rutan Aircraft Factory) is located. We keep our 1952 Cessna 170B at nearby Cable Airport from which it is a 50 or 55-minute flight to Mojave. The drive, which we've never made, can take 2½ to 3 hours. The climb up the face of the rugged Sierra Madres, close to Los Angeles but barren and rocky, is a great way to start a day. These hills have taken their toll of the unwary airmen over the years.

Once you top the mountains, usually at 8500 or 9000 feet, you make a radio call to Edwards approach to find out if Restricted Area 2515 is "hot." If the famed Edwards Air Force Flight Test Center is working, you must deviate over Palmdale and on up the highway and railroad to Mojave, a WW II Marine test center. However, on most weekends, the air is free and "Eddy Approach" will supply traffic advisories and let you fly directly over this great base. At the time of the development of the first supersonic aircraft, the broad drylake at Edwards AFB was perhaps the most valuable piece of real estate in the free world. It provided isolated security plus a broad table-top airfield where very high-speed aircraft could be landed safely "gear up." The sonic barrier was first broken here by the X-1. Even the first flight of the NASA space shuttle was made here.

Some of these flying visits to Mojave have been a bit stimulating. Winds swirling over the Sierra Madres can cause their own special brand of turbulence where you have the camera cases and yourselves well tied down in flight. Midday summer dust devils and blossoming cumulus over the desert make for excellent sailplane flying, but do produce a rough ride in a light powered aircraft. When you sit in the air-conditioned sanctity of Rutan's back office and design room, you frequently hear the wind pick up, bang doors, rattle the roof, blow sand, and generally make its presence known. You never leave your airplane, particularly a taildragger like ours, parked in front of Mojave's RAF without having it securely tied down. On more than one occasion, our departure takeoff has been made right from Rutan's tiedowns, facing down the taxi ramp into a 30-35 knot wind that would "beep" the stall warner even before the plane was untied. Under these conditions, one or more of Rutan's staff would hold the wing struts until we had completed our run-up and were pointed directly into the wind. These are just some of the interesting things that happen at Mojave.

It is certainly fitting that Rutan's developments have been made in the Mojave Desert, "over the hill" from the Southern California hub of aerospace activity. This is where many aviation and space breakthroughs have been made. In years to come, it will be most interesting to see if Burt Rutan's developments in canard designs are not among the most important engineering and flight test developments to come out of this location, the cradle of modern aviation.

Don & Julia Downie

Foreword

In every field of creative activity the influence of one person will, at some point, change the course of that activity dramatically for all time.

Burt Rutan's influence on aviation is profound, and it is interesting to take a look at this phenomenon.

Burt's productivity is legendary. His energy and drive have made it possible for him to complete projects in weeks that others might expect to require years.

Originality is, of course, the hallmark of Burt Rutan. This is true in operating procedures as well as airplane design. His car-top test rig used to measure stability and control characteristics of design configurations has been used by others—but by no one more effectively than Burt.

Canard airplane configurations go back to the Wright Brothers, but none have been as successful as Burt's. Many new designs are now showing up as canards, and it is easy to speculate that Burt will be remembered as the man who re-introduced them.

Most designers create only one airplane design and the rest of their efforts are variations on that theme. It has been said that Burt doesn't even copy himself.

Integrity is another of Burt's attributes. This is manifested in his reluctance to release designs without adequate testing.

There is no doubt of Burt's influence on the future homebuilt aircraft, and it is very likely that his innovations will point the way for improved general aviation aircraft as well. All aviation is his beneficiary.

John Thorp

John Thorp is one of the most respected aircraft designers in the United States—or any other country. The developer of many innovative designs, from the pre-WW II Lockheed "Little Dipper" to his popular T-18 homebuilt, Thorp has earned the respect of the entire aviation community. —Don Downie

Chapter 1
The Homebuilt Scene

The homebuilt scene in aviation is a careful melding of innovative designs and a pioneering of new materials, systems and structures to produce some of the finest airplanes in the air. It's the builder and the pilot (not always the same person) who explore the flight test envelope on each new product, and there are many of them. It's the pilot-owner who takes his new pride and joy cross-country, spreading the gospel of his new design throughout the land. It's the FAA, in part, who watches like a fitful parent as these offspring leap off into a world of flight sometimes larger and more challenging than that visualized in the designer's eye.

The homebuilt scene also encompasses the Experimental Aircraft Association (EAA) (Fig. 1-1), the veteran organization of homebuilders who share experiences, attend regional chapter meetings, and plan local fly-ins that proliferate during good flying weather—culminating in the annual Oshkosh Fly-in (airshow, symposiums and hardware displays) where attendance figures are in the tens of thousands each August.

The homebuilt scene is a change of life for the neophyte. How about sharing your bedroom with a growing fuselage or wing section? Condo dwellers build in the living room, while those blessed with a private garage or carport have autos that have never seen the protection of shade. Chicken coops and overgrown packing crates serve as building or storage areas. There are new friends and a whole new vocabulary. Hardware is now pop rivets, epoxy and steel tubing. The wish book comes from Aircraft Spruce & Specialty Co., Ken Brock Manufacturing, Stolp Starduster, or

Fig. 1-1. The Experimental Aircraft Association (EAA) is an organization of homebuilders who design, build, and fly their own aircraft.

some other supplier of materials and parts. Sweaty palms come from awaiting the next newsletter with its list of mandatory changes or suggested improvements in construction.

The homebuilt scene can also be highly competitive. Take the case of retired fighter pilot, Don Taylor, of Hemet, California, who was the first homebuilder to fly around the world. He loaded his Thorp T-18 with 151.2 gallons of fuel for a non-stop flight from Hemet to Nassau with a return to land in Florida. Taylor flew 16 hours and 40 minutes because "he wanted to rack up one more record for my generation before Dick Rutan lays waste to the record books with the Long-EZ."

The homebuilt scene is a challenge and an opportunity to build new things and explore new horizons. It's a whole new ballgame!

AS THE FAA SEES IT

In order to qualify as a homebuilt airplane, the builder must do 51 percent of the work. This can be handled in two ways, generally. The FAA has sent an evaluation team out to visit the designer/kit producer of some of the really high-volume designs—Bensen Gyrocopter, Scorpion Helicopter, BD-5, KR-1 and 2, VariEze (Fig. 1-2), Bryan Aircraft's HP series of sailplanes, and a few others—to determine if the kit, as it is shipped to the customer, qualifies under the current homebuilt regulations. Those that did were included on a list sent to all FAA field offices, stamped "approved." All the rest of the homebuilts are evaluated on a one-by-one basis by the local GADO or EMDO inspector . . . and, yes, there are some problems. In some areas a kit can be built, but in others the builder gets a hassle out of his inspector. It boils down to the opinions of individuals. The builder usually contacts the EAA, which in turn contacts the FAA at the Washington level. In many instances, the dispute is resolved in favor of the builder.

Regarding insurance on homebuilts, some companies won't touch them; others will. Some designs with a record of numerous claims are hard to insure. Generally, premiums are rather high—a fudge factor thrown in by insurance companies to offset the many unknowns of homebuilt aircraft.

FAA and NTSB (National Transportation Safety Board) statistics on homebuilt accidents reveal some interesting facts about today's homebuilt pilot. His average age is 43.3 and his average flying time is a high 2,103 hours, with 67.7 hours in the type of aircraft he was flying when he came to the attention of these groups. While 10.4 percent of all general aviation "store-built" accidents are caused by continued flight into deteriorating weather, the homebuilder's average is down to 1.4 percent.

So today's homebuilder is a mature person; he or she has both dedication and good judgment in varying amounts.

THE EAA, TODAY AND YESTERDAY

Almost without exception, the builders and aficionados of Rutan's designs are members of the Experimental Aircraft Association.

Today the Experimental Aircraft Association (EAA) ranks as the world's largest sport aviation organization. Originally, the EAA came from modest beginnings by a small group of men in January, 1953 in Milwaukee, Wisconsin. Several aviation enthusiasts—who liked to build airplanes—got together and decided to form an organization of members with like interests. That group was led by Paul H. Poberezny, founder and current President of EAA.

Fig. 1-2. The Rutan VariEze is one homebuilt aircraft for which kits are sold. The FAA has determined that these kits meet their criteria for homebuilt aircraft.

Fig. 1-3. EAA fly-ins usually draw large crowds of aviation enthusiasts. This VariEze drew attention at a Chino, California fly-in.

When it was initially a local organization, publicity in *Mechanix Illustrated* magazine in 1955 caused a flood of inquiries from airplane lovers from all over the world asking to join. In a few years, the EAA grew from a membership of eight to a roster in the thousands. EAA's theme became the answer to the average man's desire to fly. If he couldn't afford to buy an airplane, he'd just build his own.

In the subsequent years, EAA became a strong international organization, issuing over 145,000 membership cards to people in 91 countries. Current active, paid-up membership is 62,400. EAA has also created a system of local chapters which are considered the "heart" of the organization. Over 600 groups now hold regular meetings, and many schedule their own fly-ins (Fig. 1-3 and 1-4), providing a cohesive, active force at the local level.

EAA is involved in all facets of aviation from the "grass roots" activities which bind the organization together to the complexities of Washington representation of sport aviation interests. Today, there is hardly an event in aviation in which EAA is not involved some way.

Since its formative years, EAA has held an annual Fly-In Convention. The first was staged in Milwaukee in 1953. Today they are co-sponsored with the EAA Air Museum Foundation and are held in Oshkosh, Wisconsin. The EAA Sport aviation Convention and exhibition constitutes the world's largest aviation event, attracting nearly 10,000 aircraft and 300,000 people. Between the last weekend in July and the first weekend in August air traffic at Oshkosh is three times as busy as Chicago's O'Hare Field.

No longer limited to people who build their own airplanes, EAA now represents and appeals to all sport aviation interests, including general aviation, builders of powered hang gliders, racing and rotary-wing enthusiasts. Three special interest groups have been established for those EAA members who want to focus on a specific type of aircraft and activity: the Antique/Classic Division (Fig. 1-5), the International Aerobatic Club, and the Warbirds of America.

The EAA offers its members a series of top quality publications and the opportunity to become involved locally and nationally in all kinds of aviation activities. EAA is aggressive in preserving its mutual interest in sport flying and extending it to everyone.

Eighteen years after it was founded, the EAA had grown substantially. While researching a magazine article in 1972, we contacted EAA President Paul H. Poberezny who provided the following background on the mid-stream of its development.

"Homebuilding in various parts of the country is not all the same. Homebuilding in the Los Angeles area tends to be different

Fig. 1-4. Crowds gather to inspect new homebuilt designs at every fly-in. Here a Vari-Eze is on display at Chino. Note the variety of homebuilt designs and classic aircraft in the background.

to people in those areas as compared to those in the middle of Iowa, Florida, Maine, Michigan or Ohio. I find that ideas and views are different and the reasons quite often seem to be different.

"Our 450 EAA Chapters (circa 1972) are certainly different depending on what part of the country one comes from. In Fargo, North Dakota, a man driving 250 miles to 300 miles to attend a chapter meeting one way seems to be no unusual feat, whereas in a big city for a man to drive 10 miles to be educated or entertained is a big task. Where weather is good for flying all year round, much is taken for granted. In Wisconsin, where you have only six to eight months of reasonably good flying weather, the cold winter months are spent thinking, modifying, and rebuilding while the snow is flying.

"So, our atmospheres throughout the country are a bit different; the thinking of the people, and what building means to them is certainly different. It has been apparent to me over the past 20 years in my visits throughout the country. I would say in Los Angeles area, the builder takes a bit more for granted than in other parts of the country. The great many aeronautical engineers, the availability of materials and the wealth of aviation information is probably not appreciated as much by those in the areas as it would be in Dubuque, Iowa, Burlington, Wisconsin or Laport, Indiana. In the larger aviation communities, it would mean a bit more competition as being an authority on many subjects; out in the boondocks, anyone that has a bit of knowledge would be looked upon as a real pro.

"However, all of this has been a great help in developing sport aviation.

"We don't have a great advertising program. Actually, we have been working on 'word of mouth.' While in many cases aviation publications have been getting thinner, ours is getting thicker. We do not rely on advertising to hold the organization publication together. In fact, sometimes advertising is a bit of a pain for our type of an organization because any ad that appears in the association house organ is taken by many as an endorsement of the product, which certainly cannot be our responsibility. But we certainly do our best to ensure that advertisers are ethical and have aviation in their best interests.

"Regarding accidents, I have reviewed all the accidents for the past 20 years. We are the only organization that has a complete list of accidents involving amateur-built aircraft since 1948 until the present day. I find the type of accidents that plague the amateur builder is the same that plague the general aviation pilot: proficiency, exceeding their personal limitations and/or those of the airplane, running out of fuel, buzzing or weather.

"Insurance can be obtained for an amateur-built airplane at the same price as for a factory built. National Insurance Underwriters of St. Louis, Missouri, Dave Kratz President, insured my Pober Sport back in July of 1959. It was the first homebuilt that they insured. To date (1972), they have never had to pay a claim for public liability or property damage and they insure quite a number of machines. Several insurance companies have the same statement. In fact National Insurance Underwriters will insure your homebuilt at no extra cost if you have a factory-built airplane insured with them.

"Regarding our opinion on the future of homebuilding, I find that with the great amount of leisure time on hand today, men are looking for a personal challenge. Homebuilding is growing at a rapid pace, and today we have in the amateur-built program (which includes aerobatic and racing aircraft) almost 5,000 aircraft that are completed and some 10,000 under various stages of construction. In Canada, 10 percent of all aircraft on the civil register are amateur-built. Of course, we must take into consideration that they have fewer airplanes in Canada than we do.

"I have been working some time with FAA on standards for amateur-built aircraft. In fact, as far back as May of 1959, I felt that the homebuilder should take more credit for his accomplishments through documentation of his project than he has at this time. All of us certainly don't like paperwork; however, we must be realistic enough to understand that if we are called upon by Congressmen or the public to answer for our actions, especially in time of crisis such

Fig. 1-5. Typical of an antique/classic EAA airplane is this restored Fairchild 22 flown here by Jim Dewey, Santa Paula, California.

as a spectacular accident, questions are going to be asked. We have recommended that each amateur-built aircraft have a service manual. We developed one long ago and about 10 percent of the aircraft are using them at the present time. We also feel that there should be a log for the propeller as propellers are not standard in amateur-built aircraft; they are cut down and re-twisted, rebuilt from bent propellers or handmade. The owner of the airplane and future owners should document this in the event of repair or an accident so that history and safety are perpetuated.

"I feel that our past 20 years of reasonable safety will go a long way towards working out some minimum standards that amateur-built aircraft could comply with that would waive or eliminate restrictions, especially after the airplane is out of the test flight area. We have been kicking around some new names for the amateur-built program. We feel that 'experimental' or 'amateur' are not very popular with general aviation, with the press and even with the Congressmen. And we feel that possibly after the aircraft has met certain flying requirements, (90 percent of those certificated today would comply) that we could petition for the elimination of this portion of the restriction on those aircraft that have met their 50 or 75 hours flying requirement, along with some other simple standards.

"We feel that the airplane should then move from the 'experimental category' (Fig. 1-6) into a new category called 'sport' or 'custom.' I think that the word is more palatable and would be a credit to all. And FAA is looking very favorably on this matter.

"Moving the airplane into a 'sport' or 'custom' category after the 50 or 75 hours (75 hours if you use a non-type certificated engine), could make a difference on personal insurance policies. Many policies carry the statement or exclusion that the individual is not covered while flying an experimental aircraft. This word alone is enough to take care of the legalities and most amateur-built airplanes are not experimental in the sense of the true meaning— but nothing more than a re-arrangement of some 70 years of past aviation accomplishments.

"We were extremely impressed with our great turnout at our annual convention. We thought 1972 would see about a 10 percent - 15 percent increase in attendance; however, it turned out to be a 30 percent increase—some 43,000 flight operations during the week, 13,000 people in the campsite, over 4,500 airplanes in and out of the field during the week, over 1,000 show planes in attendance. Our forums, educational lectures and educational workshops with

Fig. 1-6. EAA President Paul Poberezny feels that the name "Experimental" should be changed to either "Sport" or "Custom." The Experimental sign is displayed prominently on the inside of the canopy of the original VariEze.

actual aircraft construction covering welding, sheetmetal, and so on, are growing in popularity.

"I can't say enough for the cooperation that I have had from those that I have worked with on EAA matters. True, there are those who do not share our enthusiasm and we do sometimes get a bit impatient in trying to bring them around. It always doesn't work, but by far, the majority have been most helpful.

"Our EAA Air Museum has been doing wonderfully; we have over 125 aircraft that have been contributed. We have also purchased some 50 acres of land at a nearby airport—some 20 minutes away from our present site along a beautiful lake and river where one can land at a beautiful little airport and taxi right onto our property. It is a municipally-owned field at Burlington, Wisconsin just southwest of Milwaukee.

"We have been expanding into divisions as people who happen to love antiques, aerobatics, classic aircraft, type clubs, warbirds, or just plain enthusiasts who like to associate with each have made it evident that birds of a feather like to flock together. But unless you have an organization to provide the umbrella, the enthusiasm and interest, and unless you spread the responsibility by getting other officers, other presidents to work with, one man cannot do it.

15

Fig. 1-7. Burt Rutan talks with veteran airshow pilot Art Scholl (in checkered shirt) of Rialto, California, during the EAA's Chino Fly-in. Scholl is a veteran motion picture pilot and aerobatic flight school operator.

Financially the Association is in good hands and we have a total of some 29 employees both on our EAA and Air Museum staff (in 1972). During the summer months our part-time employment totals 58-60.

HOMEBASE FOR THE DESIGNER

Burt Rutan (Fig. 1-7) is one of a small group of pioneers essential to the homebuilt scene. He formed the RAF (Rutan Aircraft Factory) in 1969 as a part-time, one-man effort to develop non-conventional research aircraft. In 1974, RAF became a full-time business, and Burt moved into a rented WW II barracks building on the Mojave, California, Airport.

Rutan's move to the desert area of Mojave, California (Fig. 1-8), was dictated by several factors. First, he was familiar with the Mojave Desert after spending seven years as a flight-test project engineer at nearby Edwards Air Force Base. The 9600 × 300-foot runways at Mojave, a former U.S. Navy and Marine Base at the end of WW II, have 360 days of VFR weather each year (Fig. 1-9). It is far enough away from the Los Angeles megalopolis so that hangar space and housing are relatively reasonable. The 100-miles-by-road distance from Los Angeles discourages casual "tire-kickers," yet is within a short flight of all of the Southland's aviation organizations. Daily United Parcel Service helps materially.

The two VariEze prototypes (Fig. 1-10) were developed in the WW II barracks. Three years later, Rutan leased 5,550 square feet of steel hangar and air-conditioned shop and office space on the flight line.

Specialty of the house at the RAF is efficient prototype development from initial concept definition to completed flight test at minimum cost and time schedule. RAF has been entirely self-supporting—current development costs always being paid by profits from previously completed projects. The company has

Fig. 1-8. Twenty-five Vari-Ezes are lined up on the ramp in front of the RAF at Mojave. Each of these homebuilts has been inspected by FAA GADO personnel.

Fig. 1-9. The flight line at Mojave as seen from the back seat of the VariEze. Rutan's desert facility is located at the top center of the photo.

Fig. 1-10. Original VariEze and No. 1 Long-EZ are parked in front of Rutan Aircraft Factory at Mojave, California. Dick Rutan lifts the nose of the Long-EZ before locking down the nose gear for flight.

Fig. 1-11. EAA posters are prominent in the RAF office at Mojave. Pat Storch updates the bulletin board for visitors.

never accepted deposits or payments for items during the development cycle.

The RAF schedule is to work six days a week with Sunday off (Fig. 1-11). Visitors are urged to contact the RAF in advance to make certain that the entire team isn't out demonstrating at an air show.

Chapter 2
Rutan, The Designer

Just about all of Burt Rutan's life to date has been wrapped up in aviation (Fig. 2-1). He has become a pilot, an engineer, a pioneer. To learn what makes Burt Rutan tick, we explored resumes; we talked with family, friends and acquaintances; we talked with Burt. Here's what evolved.

MODEL BUILDER FIRST

Burt Rutan has never been a copier of anything. "I never knew Burt to build a model kit or work from someone else's plans," explained Burt's father, Dr. George Rutan, a dentist. "He always wanted to try something different. He worked on both free flight and radio-controlled models, but they were all original. Actually, his first efforts in the canard design were with gas-powered models."

Burt's parents, Dr. George and Irene Rutan (Fig. 2-2), live in Whittier, California, within a 2 ½-hour drive of Rutan's Aircraft Factory in Mojave. They attend many of the homebuilders' fly-ins and Irene is the avid historian for RAF. She is developing a scrapbook that she hopes will eventually contain photos of all builders of Burt's designs and their finished aircraft. Dr. and Mrs. Rutan tell that Burt has been interested in airplanes—more in design and building than flying—for as long as they can remember.

Burt won his first model airplane contest in Dinuba, California, while still in grade school. Brother Dick, five years older, was also involved with model aircraft. At the same time, Dr. George

Fig. 2-1. Burt Rutan concentrates on his drawing board at Mojave. In this recent photo, he is detailing the last of the Long-EZ drawings.

was learning to fly and eventually purchased part of a Beech Bonanza. Thus, the two brothers and their sister Nellie, now an American Airlines stewardess, were exposed to a steady diet of aviation during their formative years.

In 1959, while attending the California Polytechnic University (Cal Poly) at San Luis Obispo, California, Burt won many model aircraft competitions, including a U. S. Navy sponsored meet at Los Alamitos Naval Air Station near Los Angeles. He had a radio-controlled copy of his father's Bonanza along with several other entries. He brought back the first place trophy in the senior scale event. The following year, at age 16, Burt walked off with the Senior Control Line flying scale event with a large twin-engine model of the Fairchild F-27 "Friendship." Rutan scaled up a three-view drawing of the F-27 from a magazine, copied colors from West Coast Airlines (which later became a part of Hughes Airwest) and powered the ship with two K and B Allyn .35 cubic-inch glow plug engines. The F-27 had a complete interior,

Fig. 2-2. Irene and George Rutan compare notes during a VariEze Fly-in. Irene is the historian for the International VariEze Hospitality Club. George, a dentist, is a Bonanza pilot. Both parents have flown with Burt and Dick in all the canard designs.

hinged doors and workable flaps. The F-27 came to an untimely end when another contestant backed into its line of flight. The contestant was injured and the gas model came out second best.

Rutan's thesis for Cal Poly's B. S. in Aeronautical Engineering degree also won $500 and that year's national award from the AIAA (American Institute of Aeronautics and Astronautics). Rutan built a radio-controlled model to study differential aileron yaw and proceeded to instrument the model with on-board graph recorders. The subject of the paper prepared was "An Investigation of the Effect of Aileron Differential on Roll-Yaw Coupling During an Abrupt Aileron Roll." Dr. Werner von Braun was honored by the AIAA at the same symposium.

Since Cal Poly's Aeronautics Department didn't have an operating wind tunnel with balanced inputs at the time, Burt designed and built a unit for the 12-inch diameter section that operated at speeds up to 150 mph.

When this wind tunnel wasn't big enough, Burt designed and built a novel car-top tunnel with precise data-collecting equipment

(Fig. 2-3). This was mounted atop his 1966 Dart station wagon where test data was obtained from medium-sized models driven at 80 mph—that's as fast as the Dart would go with the equipment on the roof (Fig. 2-4). These data-collection runs were made at night when winds were low and traffic was light. It would take a 10 to 20-mile run for good data.

No, Burt was never stopped for speeding by the San Luis Obispo police, but he was pulled over several times by curious patrolmen who wanted to know what he was doing. Some questioned the height of the unit, but Rutan had designed it to be just legal.

"That car-top wind tunnel brought back just beautiful data," beamed Rutan.

Burt first learned to fly in an old Aeronca 7AC "Champ" at Dinuba, California (Fig. 2-5). His instructor was Johnny Banks, perhaps better known as a country music disc jockey. Rutan soloed after five hours and fifteen minutes of dual instruction. At the time this material was prepared, Burt has logged more than 2100 flying hours as pilot in command. He did all his own flight testing until his brother Dick joined the RAF in 1978.

After graduating from Cal Poly where he was third in his class with a 3.4 grade average, Burt had to make a decision—whether he wanted to work in space or in aircraft. Burt chose aircraft while many of his classmates went into space research, which was very

Fig. 2-3. Cartop wind tunnel mounted on 1966 Dart Station Wagon provided much early data that was not available from smaller school wind tunnels. Surprisingly, Rutan was never arrested for speeding during his 80-mph test runs (courtesy Burt Rutan).

Fig. 2-4. Another view of cartop wind tunnel (courtesy Burt Rutan).

big at the time—half-way between the first orbiting flights and the first moon landing. Burt was more interested in aircraft but didn't want to become buried in a big company like Boeing, Douglas or Lockheed. He applied at both Beech and Cessna, but felt that he didn't want to get mired down in detail designing. He eventually chose civil service work in the flight test department at Edwards Air Force Base (Fig. 2-6). This was the lowest pay and highest risk job available, but Burt felt that it would be worthwhile.

Rutan stayed at Edwards and on temporary duty with the flight test center for nearly seven years. In a personal resume he prepared, Burt capsulized his work with the Air Force this way:

"June 1965 to March 1972—Flight Test Project Engineer at the Air Force flight test center, Edwards AFB, Grade GS 12, $18,400 P. A. Conducted fifteen USAF flight test programs, ranging from large V/STOL cargo aircraft to several types of fighters. Duties included planning, coordinating support agencies, research, test conduct (including flying), data analysis and reporting. Other functions included writing flight characteristics and emergency procedures sections of aircraft flight manuals, revision of specifications and accident investigations. Supervisory experience included employee training, evaluation, and consulting in both technical and personnel areas."

His flight time during this stint included USAF copilot time, mostly during hazardous tests, in the following aircraft: UH-IN; T-37; T-38; F-4B, C, E, and Agile Eagle; F-104A & B; F-106; C-130E; C-141A; and YA-26.

One of Burt's more interesting military assignments was as Project Director of a Low Altitude Parachute Extraction System (LAPES) where he really began earning his flight pay. "You are really sticking your neck out in flight test," he commented. "Within the first three weeks I was at Edwards on the XC'142A hovering machine, we had four people killed in the ten-man group I was with. One was in an auto wreck, two were in a Beech Baron and the other was hovering the the XV5A."

One year after going to Edwards AFB, Rutan was in charge of the LAPES program that went to El Centro, California Naval Air Station, ostensibly for three weeks. The project lasted 14 months. Rutan thought it was a good assignment despite the hot desert location at El Centro because he was in charge of an

Fig. 2-5. Rutan soloed in this Aeronca 7AC Champ at Dinuba, California in five hours and 15 minutes (courtesy Burt Rutan).

Fig. 2-6. Burt Rutan, lower right, with military flight test crew in front of a Lockheed C-130. The designer chose the lowest pay and highest risk job available in his military aircraft testing (courtesy U.S. Navy).

eight-man test crew with "lots of responsibility and 150 miles away from the boss."

On at least one flight, Rutan earned his hazard pay. A 50,000-pound load was to be extracted at fifteen feet from the cargo hatch of a Lockheed C-130. The Engineer was aboard the Air Force C-130 during a LAPES in which the parachute deploy cables snapped as the 25-ton load was still partially inside the cabin (Fig. 2-7). Under this condition, the aircraft pitched up uncontrollably even with full forward stick, pulling 2.8 G's on a 3- G airframe at an altitude of 15 feet. Fortunately, the palletized load continued out of the plane and crashed into the test site, completely ruining the cargo (Fig. 2-8). The transport aircraft zoomed sharply, but became controllable again just as soon as the cargo passed the rear of the hatch.

"If that load had jammed in the aft of the cabin, we'd have pitched up and promptly back into the ground, permanently," said Burt.

THINKING AHEAD

Many families have a habit of retaining correspondence and mementos of their offspring. George and Irene Rutan are no exception. Family communications give a singular insight into Burt Rutan at this point in his career. While Burt was working for the Air Force at Edwards AFB, he kept his family ties via tape cassettes. Dr. George and Irene came across one of these old tapes during the research on this project and shared these "thoughts from Burt to Pop," transcribed from one such cassette.

. . . as to what I want to do later—right now I kind of step back and say "What am I going to do now? What am I going to be doing five years from now?" And every time I ask myself that question, without exception, I come up with: I want to really make something of myself in aviation. I want to set a really good goal. I want to make it before I get too old. I want to get there quick. I do have patience, I think, but I do want to set a goal. I don't want to set an exact goal; I just want to reach out in this direction. I want to have a good position. I want to have my own airplane that I can get in and travel. I want to go places. I don't want to go places like New York and Chicago—I want to go to places like Muskogee, Oklahoma, and meet people. I want to do some sightseeing and have some fun. I want to fly low and see the country. I'd give a lot for the freedom I will need to do all this, I guess.

One of my first goals is to get that airplane (Viggen) done (Fig. 2-9) and refine it into a workable, buildable system and use what talents I have in management and marketing together with a good common sense approach to get good publicity and sell some plans for the thing. I'd like to make it well known; get the name Rutan in the magazines; get it into a certain segment of aviation.

Fig. 2-7. Parachute risers break as a 50,000-pound load is extracted from a C-130 at the El Centro, California Naval Air Station. If this load had jammed in the cabin, the C-130 would have crashed (courtesy U.S. Navy).

You know, I used to be pretty well known with a small segment of people. That segment was the West Coast, even the National model airplane flyers. Actually if one was very lucky, he could pick up a model magazine and page through it and find a picture of Burt Rutan. And it says he made an accomplishment; he won something. When I'd go to the meets, it was, "Burt Rutan? Oh yes, he was the guy that completely revolutionized the model aircraft carrier event by coming up with an airplane that he pulled the tail around on and it went real slow and got a lot of points even though it looked ugly." And shortly everyone copied me and were doing the same thing—flying airplanes that flew real slow. I kinda came up with it first and I won something at every meet, and I was able to do something in this small group of people. I thought it did me good; I thought it was really great. Something every kid should have.

I'll never forget that poster Mom and I saw when we were in Dallas. Remember, Mom? It said "Model Aviation Builds Better Boys." I think that kids nowadays that are getting into trouble in high school and grammar school, if they were as involved in any type of hobby—anything they'd like—anything that turns them on—skiing, fishing, anything—like I was involved in model aviation, you just wouldn't have . . . or rather you would have a better America. Anyway, right now I want to get back into something like that. It made me feel really good to be in that kind of segment. Right now, I'm not. Model airplane flyers no longer know about Burt Rutan—even the ones at Lancaster.

Fig. 2-8. C-130 pitches up as 50,000-pound payload just makes it out the cargo door. The pitch-up pulled 2.8 G's on a 3-G airframe at 15 feet. Rutan was aboard on this test (courtesy U.S. Navy).

EAA is the next place that I want to become known. This is something that's just wide open! So many people in EAA just don't know what they're doing in areas where I really know it. I want to teach them. I want to give symposiums and speeches on how to design airplanes; how to build a better way; how to solve this problem or that. I think I have just the type of education and flight test experience that would be really valuable in this, and these guys would really be interested and I could feel really good in showing them a few things and getting back from them their knowledge that would help me. Sharing you know. Getting involved in a challenging segment of society. Travel around, go to their meets, show up and get in the magazines. Learn from it and improve myself.

Just working toward a relatively narrow goal, I guess it is, but with the chances of making a name and being able to expand into that. Maybe even later on making a business out of it. Lots of people do. Like Hegy. [Ray Hegy is still making propellers in Marfa, Texas, at press time.] *He's made propellers for forty years. Goes to all these meets in his little airplane. He really enjoys himself, I think. I don't want to be quite that narrow; I kind of want to stretch out and reach a major goal, though. This is an area that I think I can move fast in. If I set myself a goal like to be President of McDonnell-Douglas, I think I*

Fig. 2-9. Burt Rutan sits in the half-finished cockpit of his prototype VariViggen in a garage at Mojave. He was working for the Air Force as a civilian engineer at the time (courtesy Burt Rutan).

would get bored before I got there. But I want to move up in the aviation area.

Gerry and I got the F-4E into a surprisingly vicious spin yesterday, but he popped out the drag chute and we recovered. There are rumors around the base as to what we did. I've been working on the data all day today . . ."

MILITARY FLIGHT TESTING

Later at Mojave, Rutan was project engineer on a series of spin test programs with the McDonnell-Douglas F-4. Following completion of this test program, Rutan made a presentation to the Aerospace profession at the 14th annual convention of the Society of Experimental Test Pilots. In this report he detailed some of the phases of the project which were to earn him, as a civilian, an Air Medal from the Government. Excerpts from his presentation which was entitled "Fighter Testing-Spin Test or Spin Prevention Test?" follow:

"Let's look at what a contractor or other test agency is faced with when it comes to the spin program. He must satisfy specification requirements and will do so, of course, at minimum expense or with a minimum number of flights. Often he is required *only* to demonstrate recovery after a specified number of turns with several spin entry conditions and possibly one or more external store loadings. Thus, his test planning task is narrowed down to a safe progression of tests building up to the goal—'demonstrating five turns with a recovery within two additional turns.' Once that goal is completed he's done. The general approach is to hold pro-spin controls for the required number of turns, then swap to the predicted optimum antispin controls.

"I submit that the most important aspect of high angle of attack handling qualities is too often being overlooked and that is *spin avoidance* or *spin resistance*. Maneuvers during the stall tests are generally terminated at adequate warning or maximum usable lift. Subsequent spin tests are conducted by doing intentional spins to meet specifications. Insufficient objective requirements are applied to guarantee testing will be accomplished for adequate spin resistance during tactical operations.

"One might ask why a prompt spin recovery requirement is not sufficient. Several reasons are pertinent. Often the out-of-control maneuver or spin mode entered from a tactical entry with immediate recovery attempts is significantly different from the mode experienced following an intentional spin entry with pro-spin

Fig. 2-10. Flight test crew with the U.S. Air Force F-4E at Edwards AFB before the aircraft was lost. Rutan is second from the right in this U.S. Air Force photo.

controls. Recovery characteristics (or success) from one mode cannot always be applied to another. Particularly with fighter/bombers, sufficient altitude is not always available for recovery from any spin; therefore, spin *susceptibility* and spin *prevention* information is far more important than spin recovery information. If a fighter is susceptible to departure or spins, the pilot will be reluctant to maneuver at maximum performance. Thus, its operational effectiveness as well as safety is compromised. Ignoring spin susceptibility to concentrate on spin recovery is just as ridiculous as ignoring spin recovery to design the aircraft strong enough to survive the impending crash.

"A test program was recently completed on the F-4E (Fig. 2-10) aimed primarily at determining departure and spin susceptibility and spin prevention methods. Classic spin tests had been conducted earlier by the contractor and the Navy on the F-4B, but no tests were flown with external stores or at aft cg positions.

"For this F-4E program, Major Jerry Gentry was the project pilot. Lieutenant McElroy and myself shared engineering responsibilities and the back seat for the test missions.

"The test aircraft was the second production F-4E. The E differs from the earlier F-4C & D by a longer nose housing an internal cannon, improved radar, slotted stabilator, more powerful engines, an additional fuselage fuel tank and fixed inboard leading edge flaps.

"Special modifications included a mortar-deployed 33 ½-foot diameter spin recovery parachute housed in additional structure built on the tail cone (Fig. 2-11). The production drag chute was located below the spin chute. Packed with the big spin recovery chute was enough riser to place its canopy 110 feet aft of the aircraft when deployed. This long riser length was determined from a McDonnell-Douglas development program after a shorter riser system failed to effect a recovery from a flat spin during the Navy spin program. External stiffeners were installed on the aft fuselage to absorb loads from the spin recovery parachute. Onboard cameras were mounted over the pilot's shoulders and on top of the fuselage to document the pilot's view of aircraft motions.

"One of the more interesting findings of this program was that the F-4 will depart and spin without any aileron or rudder inputs. This is due to directional instability above 22 degrees angle of attack. Aileron inputs usually determine the departure direction (left departure with right aileron); but the susceptibility to depart and to spin is not significantly increased by aileron inputs if angle of attack is not increased. Control surface misrigging, stability augmentation malfunctions, and out-of-trim conditions also have no significant effect on departure or spin susceptibility.

"Excluding the nose-high, low speed zoom or tail-slide type of stall entry, any severe out-of-control event was either a rolling departure or a spin.

"The rolling departure was the most prevalent and was characterized by an initial divergency in yaw followed by a rapid roll. Most were only one roll but a few went two or three. They were easily distinguishable from spins by their lack of a sustained yawing motion. Stick smoothly full forward always effected recovery. Forward stick and the drag chute effected a more rapid recovery.

"Of the many departures experienced in the test program, 101 resulted in spins. Let me emphasize that these were not intentional spins in the classical sense. The aircraft was forced to excessive angle of attack, often abruptly where it departed, but no lateral-directional inputs were applied with the intention of producing a spin. All spins developed shortly after departure without going through the rolling departure phase.

"Two flat spins were experienced during the test program. The first developed during a high angle of attack, highly oscillatory spin with an asymmetric load when the stick-aft recovery control procedure was being investigated. Once the flat spin developed, the stick was returned to the forward stop but an aerodynamic recovery could not be attained. The aircraft was recovered with the spin recovery chute after 17 ½ turns." (Burt was aboard on this one.)

"The second flat spin developed shortly after departure from a nose-low transonic entry with a clean loading. Again, an aerodynamic recovery could not be attained and the spin recovery chute was deployed. The spin chute separated from the aircraft before blossoming due to a failure of the attachment mechanism. The aircraft continued in the flat spin, the crew successfully ejected and the aircraft was destroyed.

"In summary, this test program revealed that a new spin recovery control procedure was successful for all external store loadings, was easy for a pilot to apply and did not require critical recovery timing. New information was obtained for the Flight Characteristics section of F-4 Flight Manuals. A formal training film was prepared for F-4 aircrews reflecting the findings of the test program."

Later Rutan reported that he was "very happy" with the parachute system after the 17-turn recovery. "I thought that we'd never lose the plane after that. We had planned to be more conservative on the next few flights, but lost the aircraft on the very next trip that was not supposed to be very hazardous. Project pilot Major Jerry Gentry and Lt. McElroy bailed out safely."

At the conclusion of this program, Rutan was presented the Air Medal with the following commendation:

Mr. Elbert L. Rutan distinguished himself by meritorious achievement while participating in sustained aerial flight as project engineer, Performance and Flying Qualities Branch, 6512 Test Group, at Edwards Air Force Base, California, from 1 October 1969 to 7 August 1970. During this period, Mr. Rutan demonstrated outstanding airmanship and professionalism in accomplishing his duties as a project engineer on the F-4E Stall/Near Stall Test Missions, exploring an unknown and hazardous portion of the flight

Fig. 2-11. U.S. Air Force F-4E in flight at Edwards AFB during spin test programs. Note spin chute parachute housing added at the tail of the airplane. Rutan flew many of the test missions in this aircraft. It was destroyed by a parachute malfunction during the program, but both pilot and observer escaped via parachute (courtesy U.S. Air Force).

Fig. 2-12. Burt designed and began work on the VariViggen while still working for the Air Force at Mojave.

envelope. The professional ability and outstanding aerial accomplishments of Mr. Rutan reflect great credit on himself and the United Stated Air Force.

WHY THE CANARD DESIGN?

Rutan designed and partially built the prototype canard VariViggen (Fig. 2-12) while working for the Air Force at Edwards AFB. Why the canard design (Fig. 2-13)?

Rutan had built his first radio-controlled canard design back in 1964 when he was just finishing his junior year at Cal Poly. At that time, the Swedish SAAB Model 37 canard fighter (Fig. 2-14) had not flown. The SAAB replacement for Sweden's "Draken" family of aircraft had began in 1962. The double-delta design (Fig. 2-15) combined STOL (short field takeoff and landing) with Mach 2 performance for Sweden where ordinary roads served as landing strips under combat conditions (Fig. 2-16).

The designer liked what he saw in the initial Swedish "Viggen" and set about to design something in the same ballpark. Candidly, he explained (perhaps for the first time), "I wanted an aircraft for myself as close to a modern fighter as possible—something like the F-104 or F-4. I wanted a big stick, an array of buttons, high rate of roll—a real 'macho machine' where I'd really feel like I was flying a century-series fighter."

"I went ahead with the basic natural stall-limiting function of the canard configuration. The Viggen design (Fig. 2-17) gave me a test bed to explore the natural angle-of-attack-limiting, stall-proof concept that we've followed throughout all subsequent designs."

The engineer admits that he really didn't plan to leave his long tenure at Edwards AFB, but when he was approached by homebuilt designer-promoter Jim Bede to work for him, Rutan journeyed to

Bede's Newton, Kansas facility twice before deciding to move—and even that decision was on a short-term basis.

Rutan had advanced rapidly to a GS-12 rating at Edwards. His next promotion would be to that of Section Chief where he would do no flying, but would attend many meetings and do program evaluations. "I liked my job better than those my bosses had. There were excellent benefits with 20 days of vacation a year, but I could take a leave and return to Edwards anytime within three years and retain all these benefits.

"When I finally decided to go with Bede, it was just like taking a sabbatical leave from my regular job. At that time I knew I couldn't quit the security of the civil service job, so when I visited Bede in Kansas for the second time, I planned to stay for only one year, and that was primarily because somebody wanted to pay me to work on homebuilt aircraft."

TEAMED WITH JIM BEDE

When Rutan moved to Valley Center, Kansas to work on the Bede project, his nearly-completed VariViggen was shipped right along with the household furniture (Fig. 2-18).

Rutan's job at Bede was described as follows in his resume: *"Director of the Bede Test Center, Bede Aircraft, Newton, Kansas. Basic salary $18,500 P.A. Have directed development of three aircraft types including all flight and ground tests. Had total design and development responsibility for the jet BD-5J and trainer/*

Fig. 2-13. Burt tries out seating in partially completed VariViggen. Note fighter-like visibility from cockpit.

Fig. 2-14. SAAB JA-37 Viggen. This all-weather fighter was produced for the Swedish Air Force. A total of 329 were built. The new JA-37 carries a normal armament of six air-to-air missiles plus a formidable new 30-mm cannon. It was this design that inspired Rutan to proceed with his VariViggen (courtesy SAAB-SCANIA Aerospace Division).

Fig. 2-15. Canard configuration shows well in this high-speed photo of the Swedish SAAB JA-37 Viggen. Note small contrail streaming off the wingtip of the canard surface (courtesy SAAB-SCANIA Aerospace Division).

simulator. Perform administrative and managerial duties for test department employees including four engineers, nine experimental mechanics, one buyer and one secretary. Assist marketing with demonstrations, seminars and technical presentations."

Part of Rutan's agreement with Jim Bede was that he could continue to develop his VariViggen and sell plans to it without having a conflict of interest with his production job.

On a very cold Christmas week in Kansas, Burt Rutan wrote this letter to his parents who, as some parents will do, kept it and supplied it to us. It makes interesting reading in that it foretells really what Rutan wanted to do with this portion of his life.

Dear Mom and Pop, Here I am again working on the plans. I realize I haven't written you in quite awhile. I've gotten about 100 orders for the plans now and still don't have them completed, so Carolyn and I have been really burning the midnight oil making out the final layout and inking, so we can satisfy the many people who have paid for their plans. They cost $51— nowadays. I don't think I'll have to worry about the IRS challenging the existence of RAF since it looks like we will show a couple thous. $ profit this year. Hope to sell more when we advertise! The plans are being done up first class with total of about 63 pages, completely detailed. I found I couldn't do them simple like I originally planned. Once I got started detailing things, I found I just couldn't leave things out. It's quite a job; about 400 hours total, but I'm having 500 copies printed, so if we can sell them it will be well worthwhile. I'm having them collated and mounted in plastic ring binders; printing and binding will cost about a thous. $. I'll send you a couple of copies as soon as we get them out of the printer—in about 3 weeks!!

Fig. 2-16. The SAAB Viggen is designed to take off and land in distances as short as 500 meters. This STOL performance enables the Viggen to operate from damaged runways, small secondary airfields and stretches of road. It can be supersonic within 60 seconds of the start of takeoff. The dual landing gear is designed to withstand carrier-type landings. Rutan's VariViggen was inspired by this Swedish fighter design (courtesy SAAB-SCANIA Aerospace Division).

Fig. 2-17. The VariViggen takes form while Rutan spends his days in hazardous test flying at Edwards AFB (courtesy Burt Rutan).

We had our Christmas alone at home, just the family. Not a lot of hassle and just ended up being just a short break in drawing of the plans.

"Things are about the same at work. Seem to keep busy with all the little projects and flying the jet a lot. Haven't touched the Viggen since I've been so busy on the plans.

It sure has been hectic lately; I would sure like to leave it for awhile and come on out and relax at the Hemet place, but I doubt if that's going to happen soon. Oh well, maybe some day RAF will be big enough so I won't have to report to someone and be my own boss; but guess I'm just dreaming there, too. Well, I guess I'm itching to get started on another airplane of my own. Got a lot of ideas but haven't decided yet whether it will be a little single place or a four-place made for cross-country. I'm confident I can build a 175 mph four-place with over 26 miles per gallon. What with the 55 mph speed limits that would be an extremely attractive way to travel with the gas shortage. It would be considerably simpler to build than the VariViggen actually, but then I'm not too far along on the design.

Sure would be nice to be able to work on something like that full time. I guess I will someday since I believe like you do: that a guy should do what he enjoys as much as possible.

After just over two years with Jim Bede, Rutan decided to go to work on his own. Bede and Rutan parted on very good terms. "I was against the decision to try to certify the Bede aircraft, but at the time I felt that Jim might be able to pull it off. When I decided to

Fig. 2-18. Rutan tries the propeller for size even before the engine is installed in his VariViggen (courtesy Burt Rutan).

leave, Jim and the factory threw a big party for me. It was all very interesting."

At this time, Rutan was already well into selling plans and developing support for the Viggen design. Originally Rutan had envisioned making a very simple, skimpy set of plans for experienced builders who liked the canard design to follow. "These were supposed to sell for $27 in those days," explained Burt. "However, it didn't work out that way. Everytime I stopped drafting, there was something else to detail." The first sets of plans cost $51.

After six months of work on the Viggen, Rutan was making enough from the sales on plans, cowlings and machine parts to strike out on his own.

Rutan's original VariViggen, now in the EAA Museum, was painted to copy the Air Force Thunderbird's aerobatic display team (Fig. 2-19). Rutan had flown with them as an observer. The unique pusher canard configuration soon drew the nickname of "Thunder Chicken." This name was eventually painted on the side of the airplane.

The prototype VariViggen N27VV (Fig. 2-20) won the coveted EAA award for Outstanding New Design at the 1974 Oshkosh fly-in. This was but the first of a number of EAA honors for new design.

A HOME AT MOJAVE

Rutan returned to California and to the thing he really wanted to do: design his own airplanes and do consulting assignments for aerospace companies in advanced engineering. It was a gamble, one that many engineers have wanted to take all their lives.

Rutan picked Mojave, a windswept spot in California's Mojave Desert. The sprawling former military base has large runways (Fig. 2-21), WW II Marine Corps barracks, a few aging hangars and is a boneyard for large outdated aircraft (Fig. 2-22). It is also a haven for sport aviation enthusiasts who are required to fly 25 or 50 hours over "unpopulated areas" before receiving FAA approval on their new homebuilt models.

Mojave qualifies as "unpopulated," despite claims of the Chamber of Commerce. When asked how many people lived in Mojave, Rutan quipped, "About half of them. The population is about the same as the altitude, 2787."

Most of the residents of Mojave work on the railroad. The town is a staging area for freight trains going over the Tehachapi Pass between Los Angeles and Bakersfield, the high point on the west coast rail route.

In his second newsletter, then called the *VariViggen News*, Rutan advised his slowly growing number of builders that:

"We are now conducting a full-time business primarily to support VariViggen builders. Our facility on the Mojave Airport (100 yards S.E. of tower building) consists of an office and shop sufficient to allow us to provide VariViggen components, related

Fig. 2-19. Rutan's VariViggen in flight with a new paint job and engine cowling installed. Rutan reported, "An F-106 pilot who flew it said it handled more like a F-106 than any type he had flown, military included" (courtesy Don Dwiggins).

Fig. 2-20. The VariViggen in its configuration for initial test flights near Newton, Kansas. Notice tip fins and temporary cowling. After flight tests proved that the tip fins were unnecessary for stability, they were removed (courtesy Burt Rutan).

engineering support for VariViggen builders, technical and educational material (the car-top wind tunnel project is aimed primarily at high schools and colleges), and engineering analysis/test consulting.

"We now see an important need for a periodic newsletter, complete enough to give all the information to builders that can assist them in their projects. Future newsletters will include essentially the same format information and photos as this one, with more builder-submitted information as it becomes available. All suggestions are considered. Remember this is your newsletter."

Even today, Rutan figures that he spends 15 percent of his time on his newsletter and technical reports. He writes the original material in longhand on a legal pad. Then Sally Melvill types up the copy and Burt does the layout and penned drawings.

The rented WW II barracks at the Mojave Airport lasted for almost three years before the present 5500-sq. ft. steel building was erected on the flight line (Fig. 2-23). This ramp-side air-conditioned structure provides immediate access to the 9600-foot hard surface runways at Mojave. Rutan notes that the Mojave Desert provides an ideal atmosphere for flight tests—360 VFR days per year and a low population. He didn't mention it, but high desert habitues know that the winter mornings can be bitter cold and high winds will blow sand into every corner of a building or an airplane, but Southern California smog it has not.

Burt has assembled the essential hardware needed to do his engineering job. There is an Apple II computer, 48 byte with disk

41

and tape storage using color video and Centronics printer output for engineering computations and the storage of mailing lists. A color video camera with portable video cassette recorder is used to document critical flight tests and display this data to visitors.

Since Burt lives within a couple of miles of the airport and the RAF has a Grumman Tiger as well as the twin-engine Defiant, the prototype Long-EZ and at least one VariEze, Burt really doesn't do a great deal of car driving. However, he did buy a nostalgia car—a 1950 Ford from a used car lot in Lancaster, similar to the car that his Dad owned while he was graduating from dental school. Burt paid $2600 for the classic; that's $900 more than it cost new, but the two-door sedan had only 53,000 miles on it.

"I had my first date in high school with the Ford my Dad owned," reminisced Burt. "The family kept that car for all the years when I was growing up and I have some pleasant memories to go with it. That's why this new-to-me old Ford is something special."

In an RAF facility brochure, Rutan notes that his small company has four full-time employees. Burt lists himself as "owner, part-time airplane designer." In general RAF excels in efficient prototype development, from initial concept to completed flight testing at minimum cost and time schedule. RAF has been entirely self-supporting, current development costs always being paid by profits from previous completed projects. RAF has never accepted deposits nor payments for items during the development cycle.

A VISIT DOWN-UNDER

Burt has traveled over much of the free world giving talks and seminars on this country; he has journeyed to Canada to aid in

Fig. 2-21. Mojave Airport as seen on final approach from the author's Cessna 170B. Rutan's factory is located at the far right of the flight line just beyond the first large hangar.

Fig. 2-22. Rutan Aircraft Factory with the Mojave Airport in the background. Rutan's new building is the white-roofed structure in the center with a line of visiting VariEzes parked outside.

approval of his novel design concepts. He has traveled to England, France and Germany for seminars. Most recently, he visited Australia and New Zealand at the invitation of the sport aviation enthusiasts down under. In a brief two-week trip, he was able to conduct seminars in Auckland and Wellington. Both Australian and New Zealand Governments provide a composite aircraft construction school to teach methods and inspections. Some of the class projects include building a VariEze, confidence samples and bookends.

A report in *Sport Flying* by G. Murphy who helped arrange the visit listed the following highlights:

"A small welcoming committee met Burt at 0715 hours at Auckland airport on Saturday. Burt was driven to his host, Alistair McLachlan's place where we had a brief rundown of the morrows Seminar. After a fascinating hour's chat, we left Burt to Alistair and Ian Williams with Saturday uncomitted to get the guy some rest.

"You can't keep a good man down though, and later found Burt seeing Auckland from the right-hand seat of Ian's Model D11. A trip, Ardmore/Mangere/Dairy Flat/Ardmore, left Burt agog after the Mojave Desert.

"What an experience Sunday was for 65 people! Additionally, it was very satisfying to have committee, allocated their duties only seven days before, make such an event click into place so smoothly.

"The team arrived at 0830 hours, and morning tea provided an excellent venue for people to come together before the seminar.

"We kicked off at 1000 hours, and it was soon evident that Burt's willingness to talk so capably, plus the audience's unstinting interest, was going to make our timetable difficult to meet.

"The morning presentation covered the background and history of Burt's various designs and a summary of their capabilities with a very good overview of the aerodynamics of canards. Burt's talk was accompanied by slides and a mind-boggling 16-minute movie.

"From the beginning Burt captured the audience's attention by showing a color movie of the Defiant, his new five-seater canard twin aircraft and discussed its salient features and remarkable performance. He then described, with the aid of color slides, his VariEze, Long-EZ and Quickie designs. He showed, with the aid of design and performance curves, how his high aspect ratio, high lift canard designs, coupled with his moldless composite foam/fiberglass construction method are indeed superior in performance, economy, safety and stability as compared with conventional aircraft.

"During the presentation it was immediately apparent that Burt's philosophy of 'The Ultimate in Sophistication is Simplification' and his approach of disregarding convention and going back to basics was inherent in all his work.

"We overran to 6:00 p.m. and could have gone on for another day I am sure. A measure of Burt's ability and enthusiasm is that all day not one yawn was stifled, nor was one foot shuffled."

Fig. 2-23. This sign is on the glass sliding door between the front office of the RAF and the hangar-shop area. 'Nuf said!

Rutan is completely dedicated to the task of designing and refinement (Fig. 2-24). He would be the first to admit that he is not a "people person," but some of his critics have been a bit harsh. One British journalist commented, in print, "Some of the visitors are apt to find that Rutan, an intense 36-year-old (at that time), is too busy to talk to them. 'Look around but don't touch anything,' he says, scarcely glancing up from his drawing board . . ."

The British reporter continued to quote a VariEze builder who said, "We had come a very long way, about 6,000 miles, to see his project. We spent about an hour with him, then we pushed off, we couldn't stand it anymore. We didn't feel we got a very good response, quite honestly. In fact, we were steaming about it for 11 hours on the flight all the way home.

Fig. 2-24. Burt at the controls of the Long-EZ in front of the RAF hangar at Mojave prior to a demonstration flight. World War II aircraft in the background are but a few of the unusual aircraft to be found at Mojave.

Fig. 2-25. Rutan clowns it up with a balsa wood model glider modified with a penknife to a canard design. The Grand Canyon river raft poster in the background is a reminder for Burt of an exciting surface trip down the Colorado River.

"An awful lot of people think that Rutan is God, but we altered our opinion of him. As far as we're concerned, his aeroplane flies absolutely beautifully. As a designer he may be great but as a public relations man he's pretty useless."

Rutan is frugal with the time he spends working with people vs. working at the drawing board. "We could spend all our time standing around and talking and never get any work done," he commented honestly. However, the designer appreciates that personal contacts are an essential part of a successful business, and he holds regular Saturday discussions on composite construction and demonstration flights of the various models available at Mojave. These Saturday sessions are scheduled except for the weekends when important fly-ins are listed far in advance (Fig. 2-25).

On a recent Saturday, all the other troops were out of town and Burt handled the program by himself. The front office was crowded with visitors who watched movies, viewed flight test video tapes, asked endless questions and then watched as Burt put on a flight demonstration in the Long-EZ. Normally these chores now fall on brother Dick or Mike Melvill, who is officially listed as Customer Relations man, but both were out of town.

Burt is essentially a shy person, belied by his 6'4" frame and easy-going manner. When he puts lines on drafting paper and massages flight test data, his eyes glint and he really becomes alive. The people who build his products are delighted to have it work out this way.

DICK RUTAN, TEST PILOT IN THE FAMILY

Dick Rutan (Fig. 2-26) is five years older than his brother Burt, has been involved in aviation all his life, and fits perfectly in the scheme of things at Mojave. Dick commented that their mother thought that both boys had been born with av/gas in their veins.

Dick took flying lessons while going to school in Dinuba, California, but couldn't solo because he was too young. Shortly thereafter, his father took lessons and went on to get his private pilot certificate.

When Dick was old enough, he applied to the Air Force for aviation cadet training. Since no pilot slots were open, Dick joined the Air Force and went on to become a commissioned navigator. He flew the back seat of fighters for seven years before an opening occurred to get into pilot training. He was a radar operator, a "scope dope," in F-89s and F-101s.

He flew the last actual F-89D "Scramble" from Keflavik AFB as a radar observer. After years of trying, he was accepted for Air Force pilot training in 1966 and went through Class 67B.

Dick is a firm believer in parachutes and always watches the one he's going to ride during repack. He's been forced to bail out twice in his 20 years on active military duty. Once was over North Vietnam when his F-100 was shot up on a strafing run. He took a hit 20 miles inland from the water and the F-100 burned all the way to the shore. He ejected just off the coast in the Gulf of Tonkin and spent three hours in warm, clear water before being picked up by a jolly green heli-rescue team.

Dick volunteered for three special mission tours during the year he spent in Vietnam. He was flying a Commando Saber operation for pilots in that part of the war.

"After being shot down, I was eligible to return to the States," Dick remembers. "I had visions of taking my little girl to the zoo in a couple of weeks and all those good things you look forward to on return from overseas. I was so glad my tour was over, I even went to sleep in the helicopter on my way back to Da Nang Airbase. However, the front-line fighter operation was a challenge—that's where the action was, and that's where I wanted to be."

Dick's other bailout was over England during a 3 ½-year tour with F-100s. He had been on an engineering test hop over the North Sea after taking off from the USAF base at Lakenheath Air Force Base. Weather at the time was typical of England in May, 1970—900 feet with the tops at 35,000'. Dick made his routine list

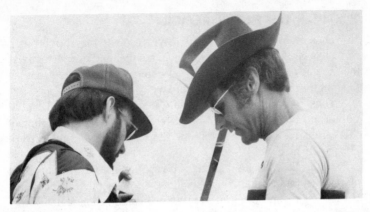

Fig. 2-26. Dick Rutan with a colorful hat, suited for all day in the sun at this fly-in at Chino, California, talks with a prospective builder.

of checks on the engineering flight; routine that is until he made the prescribed – 1 (negative) G maneuver. A mechanic had left an oil sample bottle loose and with the negative G maneuver the bottle stuck in the oil intake. Instantly there was no oil pressure, but the engine continued to run for 12 ½ minutes.

"I was half-way down the GCA approach at Lakenheath and just breaking out of the overcast when the engine literally blew

Fig. 2-27. Dick Rutan leans over the cockpit of the Long-EZ before going on a checkout flight with A.L. Letcher of Mojave.

up," he remembers. "I just had time to level out and eject. The 'chute opened and swung only twice before I was in trees. It was a nice, soft landing."

Dick has been a licensed flight instructor since 1969 (Fig. 2-27). For ten years before leaving the service, he taught in several USAF Aero Club in Cessnas and "whatever." He held the "E" (engine) part of his A&P before joining the service and completed the airframe section just before leaving the service.

At the time of this writing, Dick has a total of 6,000 flying hours in just about every type of aircraft ranging from the tiny 240-pound (empty weight) "Quickie" to the USAF C-124 transport with a gross weight of 185,000 pounds. Ask him what is his favorite fun flying machine he grins easily, "The company's first-line aircraft—the Long-EZ (Fig. 2-28)."

BURT LOOKS AHEAD IN AVIATION

As Burt Rutan looks to the future, he sees a continuing development of aircraft design. He has drawn the plans for both a Formula I single-seat racer and a biplane racer for the professional competition market. Both are being built as this manuscript is completed.

In a completely different phase of aviation, Rutan has high hopes for the future of the Predator (Fig. 2-29) and feels that it will have a significant effect on feeding much of the world. The Predator is a large, turboprop agricultural aircraft. Its configuration is a unique joined-wing strutless biplane arrangement. It uses a

Fig. 2-28. It gets hot in the desert, so Dick Rutan rides out in the air while taxiing to and from the flight line with the Long-EZ during a check ride with A.L. Letcher.

PT-6A-34 engine and has a hopper payload of 6700 pounds. It features the following performance improvements over current Ag aircraft: payload (59 percent), swath width (58 percent), speed (9 percent), climb (11 percent), stall speed (10 percent). It also promises major improvements in safety (stall-proof and crash protection). The Predator was designed during a RAF feasibility study conducted for an independent contractor. NASA has expressed an interest in the design and is currently building models for wind-tunnel tests. While Burt did the original design concept on the Predator, he is quick to credit inventor Dr. Julian Wolkovitch of Palo Alto, California, with his suggestion of joining the wings together.

Basic research is an area that Burt enjoys and does with aplomb. The AD-1 program is an intriguing example of this type of effort.

The AD-1 is a twin-engine jet research aircraft featuring an adjustable skew-wing (Fig. 2-30). It began as an RAF-submitted,

Fig. 2-29. Scale model of the Predator design shows several unique concepts in this large agricultural aircraft (courtesy NASA).

Fig. 2-30. NASA composite photo of first flights by the AD-1 at Edwards AFB. This project began as an unsolicited design proposal by RAF to NASA. The flying testbed was constructed by Ames Industrial Corp, Bohemia, New York (courtesy NASA).

unsolicited proposal to NASA for a feasibility study in December, 1975. At the time, skeptics abounded within both NASA and the aircraft industry as to the possibility of building a manned, jet skew-wing research aircraft for less than several million dollars. The AD-1 design task, accomplished by RAF between May, 1976 and February, 1977, cost NASA only $15,000. Fabrication of the aircraft by Ames Industrial Corporation, including consulting, documentation, static load-testing, and delivery to the government in a completed, flight-ready condition was accomplished between November, 1977 and February, 1979, and cost NASA less than $240,000. It is interesting to note that the above work by RAF and AIC was done at a profit and at far less cost to the taxpayer than the NASA tasks of the wind-tunnel test, simulation and contractor monitoring. The AD-1 is all composite, using a glass-foam sandwich for all basic structure. It is powered by two TRS-18 microturbo turbojets. Its wing skews 60° actuated by redundant electric motors. NASA installed instrumentation and is currently test-flying the AD-1.

In the non-aviation field, Rutan has developed a solar water heating system (Fig. 2-31) that really thrives in the hot desert of Mojave. The RAF solar water heater is a system intended to be built by the hobbyist to provide a large percentage of his home water heating energy requirements. It was designed to minimize

Fig. 2-31. Not all of Rutan's designs will fly. Here Burt adjusts the hot water flow for his solar water heater. The solar system has supplied all the hot water needed at RAF since 1977.

the cost-per-BTU ratio. It uses an east-west oriented parabolic collector and an all-composite storage tank. This system has continuously provided 100 percent of the hot water requirements at RAF since July, 1977. Burt plans to market plans for this system.

It is fully predictable that Rutan will come out with many more break-throughs in aviation. His track record is excellent. He doesn't talk about projects before they are fully developed and hopefully fully test flown. So, when you ask Burt Rutan, "What next?" he smiles, his eyes twinkle, and he says: "People would climb all over me if any information got out. Let's just wait and see."

Chapter 3

VariViggen—First Of Rutan's Canards

Much of the early reporting on the VariViggen project was written first-person by Burt. However, an impersonal report on the roll-out of the original canard aircraft was penned by Art Stockel, Technical Editor of the Air Force Flight Test Center at Edwards Air Force Base and published in *Sport Aviation* magazine in May of 1972.

"VARIVIGGEN" COMPLETED

On February 27, 1972, Burt Rutan celebrated the roll-out of his original canard aircraft, the VariViggen (Fig. 3-1). This airplane is really something different. A lot of original thought and careful engineering were combined to result in such advanced features as these:

High lift at low angles of attack through the complementary vortex interaction of the forward canard surfaces and the rear-mounted wing with its reverse reflex.

Stall and spin-proof performance which retains safe flying qualities at the point of maximum attainable lift.

Elimination of adverse yaw effects by means of the fin-and-aileron geometric interface.

Minimal trim drag at high speed as a result of variable aft wing reflex.

Burt, who is a member of EAA Chapter 49 in Lancaster, California, began working on this aircraft design in 1963 while a student at Cal Poly. Being an aeronautical engineer at the Air Force Test Center nowadays, he is not exactly a greenhorn amateur.

Fig. 3-1. VariViggen in flight with temporary 12-gallon auxiliary fuel tank mounted like a drop tank below the fuselage. Paint job is copied from the Air Force Thunderbird competition team and Rutan's version earned the name "Thunder Chicken" (courtesy Burt Rutan).

The actual construction of the VariViggen in the garage of his Lancaster home required nearly four years having begun in 1968 after extensive wind-tunnel tests and model experiments dating back to 1963.

"Model experiments" does not mean dropping a toy out of the bedroom window. It means building an aerodynamically true one-fifth scale model which was suspended on a specially built test rig atop a car. Quickly removable, the rig was clamped to the luggage carrier on the station wagon's roof. The rig allowed measurement of airspeed, angle of attack, lift, drag, sideslip, side force, roll moment, elevator-aileron-rudder positions, and an extra data channel which permitted measurement of hinge moment (stick force) or structural load, etc.

Ailerons and rudders and elevators were controllable by the test engineer through an illuminated control box at the right front car seat.

Mounted on a spherical bearing, the model could be shifted fore and aft to test effects of varying CG's. A tape recorder was used to make verbal recordings of data during the test runs.

The "captive" model provided a wealth of information for design of the prototype. The original configuration was actually quite different from the design which was optimized during the car tests.

The visitor's first impression of the airplane when he comes through the door is that it really fills up that two-car garage! The rear wing outboard panels were mated to the fuselage after the plane was removed from the garage. The span, with outboard wing panels removed, is only 8 feet, allowing road towing.

In designing the prototype, Burt decided against going for optimum high speed. Instead, he wanted plenty of wing area for safe, docile, low speed flying qualities. This was considered a conservative approach to development of the configuration and, if it proves as successful as expected, a high performance version will be built to define the performance capabilities of the concept.

He settled for a slab-sided fuselage and flat-bottomed wings for ease of jigging and building (Fig. 3-2). All curved surfaces have a single curvature, except the nosecone, visor/glareshield, and aft fuselage top. These few pieces with compound curvature are all made of fiberglas and Burt has retained the female molds in case more parts are needed.

The main structure (plywood) was easy to build using normal techniques. Windshield and aft canopy have only single curvature while a formed plexiglas piece from an HP-11 glider was fitted to the front canopy. Spruce was used for spars and longerons, aircraft plywood for formers, ribs, and skin. The plywood skin was covered with lightweight Ceconite and finished with dope followed by polyurethane. Most of the building time (and this will surprise no one) was spent on fittings and systems such as the retractable gear, the trim and reflex electrical systems, and the controls.

As if things weren't hard enough to get done, Burt claims he used all metal, flush riveted construction for the outboard aft wing panels to get experience in metal construction.

To keep things convenient later on, the nosecone is hinged at the top, thus exposing master brake cylinders, nose gear retraction system, batteries, landing light, angle-of-attack transducer, and

Fig. 3-2. VariViggen built and flown by Mike Melvill, was the first aircraft to be completed from Burt Rutan's original plans.

pitot-static system. A VOR antenna is installed inside the all wood canard surface.

The prototype VariViggen has a main wing span of 19 feet, a canard span of 8 feet, and a 19-foot length. There is a roomy cockpit for two pilots in tandem which incorporates modern fighter cockpit layout and affords really terrific visibility from both seats; that means both pilots can see well, not only forward but also directly up, down and rearward (both rudders and main landing gear wheels are visible from the seat).

The fully retractable tricycle landing gear is operated electrically, as are the trim bungee and wing reflex (more about this later).

A 150-hp four-cylinder Lycoming engine (which was bought used and then overhauled by Aronson Flying Service at Rosamond, California) powers the prototype. Burt had originally intended to build a single-place prototype with a 90-hp engine, but as many EAA builders can testify, things tend to grow, and a good deal on the larger engine turned up at just the crucial time.

Now a few words on the cockpit, which has separate opening canopy sections for pilot and passenger. The control stick has a gloriously functional handle with switches for electric reflex control, trim, and radio transmitting. Landing gear handle and landing light switch are just forward of the throttle—no need for fumbling around during approach to landing (Fig. 3-3). An angle-of-attack sensing system, used to select approach speed, also operates as back-up gear warning (besides the warning horn and light!) because it remains inoperative as long as the gear is retracted. An override switch permits the pilot to turn the angle-of-attack system on for gear-up maneuvering.

Upper and lower right-hand control consoles contain further items. The lower-right console carries the circuit breaker panel. Recording hourmeter, cockpit lighting switches, magneto switch, and engine starting controls are all on the upper-right console. This leaves the main instrument panel free for flight and engine instruments. The radio, when available, will be located on the left-side console.

The VariViggen weighs in empty at 900 lbs. and approximately 1500 lbs. when loaded. This results in a wing loading of 12.3 lbs/sq.ft. The ship has a power loading of 10.0 lbs/hp. (In case you're wondering, sweepback angle of the aft wing is 27° at the quarter chord.)

The rear-mounted engine drives a 70″ diameter, 70″ pitch wooden pusher propeller (Hegy) directly (no extension shaft). While it may appear that propeller ground clearance might be critical, this is definitely not the case, and Burt can show that when the airplane is fully rotated to take-off pitch, the propeller is nowhere near ground contact. The reason for this is that the prop blades track only 6″ below the wing's trailing edge. In fact, when the nose is lifted high enough to put the skids on the ground, the prop still has over 4″ of clearance.

An interesting safety feature should be pointed out here: the propeller is between the twin rudders, and anyone who wanted to blunder into the blades would have a hard time walking against the propwash!

In deciding on the unusual aircraft configuration, Burt felt that the canard feature had distinct advantages over the conventional as well as the delta-wing types. Obviously there is greater low speed lift available at low angles of attack because both surfaces contribute lift; there is more positive lift control with the canard approach, and at higher speeds trim drag and induced drag can be considerably reduced.

He was aware that many attempts had been made to reap these benefits, but most failed because of unacceptable stall characteristics, poor lateral directional stability, and "packaging" problems.

Fig. 3-3. Instrument panel on Mike Melvill's VariViggen in flight. Note landing gear handle at the far left and complete IFR instrumentation.

However, one design has been very successful; the Swedish SAAB "Viggen," a jet, is in production in two versions—attack and fighter (Fig. 3-4). The "Viggen" has Mach-2 speed capability, excellent flying qualities without the need of electronic 'black boxes' to augment stability, and it operates out of 1500-foot dirt airstrips! Burt said his VariViggen was designed about the same time as the "Viggen," the name having been added later. And, while it is not a copy, it shares some of the "Viggen's" design features, such as location of the canard surface high and in front of the main wing so the vortex from the canard will mix favorably with the main wing vortex to increase low speed lift. But Burt went one step further and used the front (canard) rather than rear control surfaces for elevator function. Consequently, a nose-up control input immediately increases lift—even before the angle of attack increases—and this extends overall maneuvering capability.

The most important reason for front elevator control, however, was that the main wing control surfaces could now be employed in a very interesting manner as a controllable reflex (as well as the necessary aileron function) (Fig. 3-5 and 3-6). For increased lift at low angles of attack, the reflex surface is turned downward, causing more lift to be required of the canard (elevator) to balance out the resulting pitch-down tendency. Therefore, elevators and reflexes all act effectively as flaps. As a result, instead of the drastic nose-up approach to landing, the Vari-

Fig. 3-4. SAAB Viggen JA-37 in battle dress shows the canard design that inspired this popular fleet of Rutan composite aircraft (courtesy SAAB SCAN-DIA).

Viggen's angle of attack is only three degrees at 70 mph (Fig. 3-7). This means that on a three-degree glidescope the fuselage is level!

The triple advantages of this low angle of attack are greatly improved visibility, improved roll and yaw flying qualities, and less power required for a go-around.

On the other hand, the controllable reflex is also used at high speed or cruise. It is set to minimize trim drag and fuselage drag regardless of CG, weight, or speed. As the aircraft is trimmed for cruise, the reflex is set to achieve a level fuselage attitude and minimum elevator drag (the Fowler-type surface is tucked into the slot).

Low speed characteristics are very safe. Wind tunnel and radio controlled model tests have shown that the plane will not stall or spin. It just becomes 'super stable' in pitch at angles of attack higher than about 15°. When full aft stick is applied the plane remains at a safe angle of attack. Sharp turns or banks can be made using only aileron, only rudder, or even with crossed controls, and the plane stays easily controllable. Burt says that a short field approach with minimum air speed (less than 50 mph) would be made with the stick trimmed and held full aft, using only the

Fig. 3-5. Tufts on the left canard of the VariEze remain flush with the wing surface during a full stall.

throttle for flightpath control! This may at first seem terrifying, but the data and radio controlled model test results indicate otherwise.

As it worked out, the VariViggen doesn't need differential aileron control to alleviate adverse yaw. The pressure difference on the wing's upper surface which is caused by aileron deflection acts directly on the vertical fins to turn the nose into the roll. Wind tunnel tests have shown that the plane should make coordinated turns even at approach speed while the pilot's feet are on the floor.

Cautious when it comes to predicting air speeds, Burt will only say that the VariViggen should cruise with a Thorp T-18 and land slower than a Cessna 150. He knows that all the theorizing, planning and building are now over and that actual flight testing is the next step.

THE VIGGEN FLIES

A year and a half later, Burt reported on his first flights in Kansas. First light of print on this memorable series of flights was in *Sport Aviation* in August, 1973, entitled "VariViggen Designer Builder Report", by Burt Rutan, and the text follows.

"About the time I had the airplane ready for taxi tests, I had decided to leave my job as project flight test engineer for the Air Force at Edwards Air Force Base in California and join Bede Aircraft, Inc., at Newton, Kansas. I found as I had suspected that developing and testing homebuilt airplanes is as much a challenge

Fig. 3-6. Only in a violent accelerated stall will the tufts break away from the surface of the VariViggen canard.

and a lot more fun than the supersonic jets! I was so anxious to get started at the Bede Test Center, I removed the wings from my newly completed VariViggen and packed it into a moving van with my furniture and made the trip to Kansas without having made even the taxi tests.

"The move was a good one; the aircraft arrived in excellent shape, and I was able to use the excellent shop facilities at the Bede Test Center and the expertise of Paul Griffin (Chief Designer) and Delmar Hostetler (Shop Foreman) to help me make the last minute adjustments to get on with taxi tests.

"About the first of April '72, I got the engine running and started low speed taxi tests. I found the airplane to be very maneuverable on the ground. Its geometry is such that it can be nosed up to within three feet of a hangar, turned and taxied away without ground assist. I can easily see the wingtips and rudders from the cockpit and due to its short and low wings, it can be taxied between and around other airplanes much easier than conventional aircraft.

"My original nose gear encountered shimmy while taxiing at only 25 mph. This was due to improper geometry and the use of a streamline type tire. Rather than spend a lot of time developing a suitable shimmy damper, I installed a BD-4 nose gear. This absolutely eliminated the shimmy, but since I had not provided any shock absorption, the gear gave a hard ride in bumps. Well, after all, this nose gear was just a temporary fix—who knew if this thing would even fly?

Fig. 3-7. VariViggen coming across the numbers at Mojave. This Viggen is flown by Mike Melvill who has put more time on the Viggen design than anyone else in the world to date.

"Another initial problem involved the engine. I had smooth operation up through about half throttle, then the engine would quit if the throttle were advanced further. I found I could get ⅔ throttle if I leaned the mixture, but that was all. I spent the next several weeks changing carburetors, air inlets, valve pushrods, ignition and just about everything I could think of.

"I had the local FAA in for the final inspection and took care of their squawks: mark *Fuel* outside the fuel lid. *Fuel shut-off—pull* on the fuel shut-off handle, and safety the shoulder harness bolts.

"After more tries to get the engine to operate properly, I convinced myself that it would run reliably up to ⅔ throttle and that's enough to fly. I set up a movie camera, run by Les Berven, our BD-5 Test Pilot, got the wife and kids out to the airport and started my first high speed taxi tests. I started at 35 knots and made successfully faster taxi runs in five-knot increments, checking controllability.

"The nose gear left the ground on the 45 knot run and I found it very easy to hold the nose off in any attitude I wanted, and it fell through gently when decelerating through 30 knots. On the next run at 50 knots, I exercised the ailerons to see if it would rock while light on the main gear with the nose up. To my surprise, I found myself slightly off the ground rocking the tires on the runway! I decided the next run would be a brief flight down the runway at about 3 feet altitude. I accelerated to 55 knots, rotated and flew down about 4,000 feet of the 7,000-foot runway at Newton at about 10 ft. altitude. The feel of all three axes was solid and smooth—and I was one happy guy! The landing was a grease job but in my relief I let the nose down hard in one of the ruts in the runway and the nose gear collapsed. A quick inspection revealed no major damage. The retraction link had buckled due to an impact load from the stiff gear, allowing the gear to partially retract. Faced with 90 minutes of daylight remaining and very still, no-wind conditions, I decided to make a quick repair and fly. Within a half hour I was on the end of the runway and setting my ⅔ throttle and lean mixture for takeoff.

"As Dan Cooney (Chief Bede Chase Pilot) maneuvered the Cessna 172 chase plane into position, I started my take-off roll. Takeoff and climb were normal and a very strange feeling came over me as I cleared the end of the runway. The air was absolutely still and there I was climbing straight ahead. I had waited a long time for this moment, but somehow it felt like I was on my first solo.

"I leveled off at 1500' AGL and performed some stability checks—static and dynamic—and pleased with the results, I proceeded to do sideslips and maneuvering turns (Fig. 3-8). I set the reflex at several positions and slowed up to full aft stick to check low speed handling. Again the aircraft felt solid, while still responsive—particularly in roll. So much for the work. I moved in to the Cessna for some pictures, then made a low pass down the runway and landed just at sunset after 50 enjoyable minutes of flying.

"The next nine weeks were spent completing initial flight tests, improving engine operation, gathering stability and performance data to compare with wind tunnel results, getting my 50-hour restriction lifted, and adding a new cowling and spinner. I solve the engine problem by shifting the carburetor air inlet to the cooling air inlet location and replacing my Midas—special exhaust system with short stacks.

"Tip fins on the wing tips were on for the first few flights for some extra directional stability just in case it would be needed. After tests showed that the amount of directional stability was more than adequate, the fins were removed.

"By Oshkosh '72 (nine weeks after first flight), I had logged 75 hours and had taken the airplane cross country to Illinois and Oklahoma. Oshkosh was the highlight though, landing at the convention and taxiing to our parking place among the other

Fig. 3-8. VariViggen in flight at Mojave as Saturday visitors look on. Original VariEze, still with the 2-cylinder Franklin engine, is in the foreground.

homebuilts was the culmination of those years of designing and building. When my wife and I left Oshkosh loaded with 80 pounds of baggage (including the Stan Dzik trophy for design contribution), we agreed that those who have the opportunity to participate in sport aviation do have more fun.

"The VariViggen spent the winter attending whatever fly-ins we could make, completing stall/spin tests and getting a new paint job and interior. After more problems with the stiff nose gear, I built a new unit which uses an air-oleo strut, the lower end still being the BD-4 assembly. This one is giving excellent service with no problems, even on sod runways.

"I have been unable to spin the aircraft during straight ahead stall entries or accelerated entries with all combinations of aileron and rudder controls. I haven't tried hammerhead entries (I'm chicken), but the radio-controlled model would not spin from hammerheads so I think I can safely say it is not spinnable.

"The thing I like most about the airplane is its roll qualities. Its low adverse yaw, high roll rate, ability to stop the bank right where you want, combined with the fighter visibility and cockpit just make it fun to fly. An F-106 pilot who flew it said it handled more like the F-106 than any type he had flown, military included. The roll rate is surprisingly high even at 50 knots. This allows you to roll 120 degrees to level at the top of a steep wing over and fly away without dishing out.

"Due to the position of the landing gear and thrust line, the nose wheel rotation speed is about ten knots above the minimum flying speed. Thus, on a full power takeoff, it is impossible to force the aircraft in the air at an unsafe speed. I generally make the take-off roll holding full aft stick. At about 53 knots, the nose comes up slowly and is easy to control at just the position I want for initial climb. There is no tendency to bobble or hunt for the initial climb angle since pitch damping is high and the aircraft is not sensitive.

"I generally fly final approach at 10 degrees angle of attack (all VariViggens have angle of attack indicators) which results in about 60 to 70 knots depending on gross weight. The speed bleed off at flare is fairly rapid without a great deal of tendency to float, even though there is a considerable ground effect. The ground effect is so great that if you want you can make a full stall type landing with touchdown as slow as 37 knots; that's right, 37 *calibrated*, not indicated. Full stop is easy within 300 feet.

"VariViggens are less affected by winds during taxi than other types of similar wing loading. In fact, due to one rudder blanking the other, you can taxi in a 40-knot direct crosswind with no tendency to weathervane.

"As I mentioned, rolls are fun. You can complete a 360° roll without altitude loss from a level flight speed as slow as 85 knots. But, the VariViggen is not an aerobatic airplane. It is strong enough, but due to its low aspect ratio (2.7), it slows down considerably during tight maneuvering. This makes vertical maneuvers such as loops very difficult. Also, of course, true snap rolls are impossible since you can't stall the main wing.

"The climb and cruise speed of the VariViggen are not particularly good considering it is a two place, retractable with 150 hp. It will cruise with a BD-4 or T-18 with equal power, but then, those are fixed gear aircraft. It may improve some when I get around to adding the 3 gear doors, but I doubt if it will be over 5 to 10 mph. Where the VariViggen performance really shines is at the low speed range."

RETURN TO THE DESERT

When Burt Rutan moved to Mojave in 1974, he brought the prototype VariViggen (N27VV) with him. In that year, his canard design won the "Outstanding New Design" trophy at the Oshkosh EAA meeting. In his second newsletter, then called *VariViggen News,* he wrote:

"Even though the VariViggen had been to California twice before, we couldn't wait after arriving at Mojave for the first chance to really demonstrate her flying ability to the multitudes out West. The next weekend was the EAA Western Fly-in at Porterville, California. The following excerpt from the Bakersfield EAA Chapter 71 newsletter written by Denny McGlothlen tells it all:

" 'The star of the show was Burt Rutan with his VariViggen. Boy, this bird really turned me on. I was out on the runway when Burt flew in the airshow and saw the VariViggen make the low speed sharp turns right at lift off—well, an airplane just isn't supposed to do such things but this one sure will. I can see that this is going to be a very much built airplane in the EAA ranks.'

"The VariViggen succeeded in awing the crowd there and also won the 'Most Popular' trophy, the '2nd Monoplane' trophy, and the 1st place cash prize for the spot landing contest. The VariViggen has won every spot landing contest it has entered. Due to the fantastic low speed maneuverability and visibility, you can use quick tight turns on short final to set up the correct height and speed for the accurate touchdown. The 2nd place winner at the

Beatrice, Nebraska contest just shook his head and said, 'That's not fair; that's not an airplane!'

"We have two more airshows and a magazine article commitment within the next two weeks. After that we plan to remove the old cowling, give the aircraft a good inspection (she now has 400 flight hours) and install the new design cowling with prop extension. When testing is complete on the cowling we will begin cowling production."

In the early days of development, the unconventional landing gear arrangement raised many an eyebrow, particularly when the canard was parked on the ramp. Burt discussed this feature in his newsletter, along with the reasoning behind the selection of 180-hp engine and his ideas on modifications on the VariViggens:

"Parking. Without pilot or copilot, the CG is very near (slightly aft of) the main wheel location. With weight on the main gear its reaction moves aft of the no-load position. Thus, when the pilot gets out, he lets the aircraft down on the aft skids. At first we were ashamed of this tail-sitting attitude and would immediately tie the nose gear to a tiedown or install an aluminum tube tripod under one skid whenever we parked. I don't do this anymore for the following reasons: 1) sitting on the skids, the center of pressure is well centered and the aircraft will take winds from *any* direction with little weathering or upsetting tendency common to the conventional parked aircraft; 2) when parked in a hangar, even a low wing aircraft will overlap all the way to the fuselage and thus a VariViggen will take up considerably less room than even smaller homebuilts; (I've put it in many hangars after the owner said 'Sorry, we're full' without even moving other airplanes!) 3) this attitude allows more convenient pre-flight inspection of fuel, oil, landing gear and pulling the prop through; 4) baggage loading, fuel and oil loading is convenient while on the tail; 5) it is very easy to pull the nose down by the canard tip, step to the ladder and get in when ready for ingress; 6) we consider it a 'status symbol'—just one other thing no other plane on the field can do! However, for an airshow, in order for people to more easily inspect the cockpit, we either tie the nose gear to a tail tiedown rope (VariViggens park backwards, too) or retract the nose wheel only (pull the main gear braker) and set it clear down on its nose. Thus, the canard is an excellent seat for four to watch the show!

"Engine Selection: Since I mentioned that I would like to have 180 hp, many have thought it was for more speed. Not true, considering 75 percent power cruise, speed would only increase 10 mph with 180 hp. The main reason would be for better rate-of-climb, particularly at high altitude. Remember, low aspect ratio means lower climb performance. A VariViggen will not climb as well as a conventional aircraft with equal cruise speed and hp/weight ratio. Those that want better high altitude climb performance and want to use a heavier engine or constant-speed prop will find the airplane tail-heavy and for that reason I have not recommended them, due to the terrible requirement for lead in the nose. There is a better solution, however, that can eliminate this problem. I used this solution when I found my partially completed airplane to be tail heavy. The original design had a shorter wingspan. I increased the span of the outboard wing panels. This moved the *allowable* CG range aft, thus solving the problem without lead. A disadvantage is a slightly reduced G-capability. If you are interested in using a constant-speed prop or heavier engine, send me the weight, length of engine and weight of the prop. I will then calculate for you the amount of extension to the wingtip, show how to make the extension and calculate the amount of reduction in allowable 'G'. This can only be done up to a point at which the control power of the canard is reduced and the overall CG range is too small. While this solution is better than lead nose weight, I still recommend 150 hp (180 hp for short airstrips or high density altitude flying) and a modern light weight wood prop.

"Modifications: As you know, it has been our policy to not be adverse toward those who want to modify the VariViggen. We have had this policy mainly in the interest of promoting education and design progress. However, we have seen some examples of modifications, even some under construction, that will result in disappointing performance and in some cases unsafe flight characteristics. In all cases those individuals designed their modifications by aesthetics and by eyeball rather than by valid engineering calculation supported with appropriate tests. In most cases, when I was able to point out the disadvantages and calculate the effect on performance and stability, the author of the change decided to stick with the plans. One builder doubled the rudder area and didn't even know that that would reduce overall directional stability due to rudder float.

"I must modify my policy to point out that I am not averse to anyone modifying the airplane that is qualified (or finds qualified help) and is willing to conduct the analysis and tests required to verify the modification before flying his aircraft. I am very averse to those who may give all the rest of us a bad image by building a

'VariViggen' that either has poor performance or contributes to an accident statistic under the name VariViggen.

"A plans-built aircraft has good utility and excellent flying qualities. Modifications that add weight, be they as subtle as extra heavy gussets everywhere or fiberglass over the wood skin—or more substantial—like 70 gallons fuel or four-place, etc., etc., can result in very disappointing climb performance at high altitudes. Our experience in flying N27VV over 400 hours in all kinds of flight conditions, runways, weather, density altitudes, etc., is very valuable, and we have found that due to the low aspect ratio (necessary for optimum low speed flying qualities) the airplane should have a lower weight-to-power ratio than conventional designs. You cannot expect to carry four people and more fuel adequately from Albuquerque in the summer unless you use at least 200 hp.

"You cannot expect the same safe flying qualities if you stretch the nose several feet for 'looks.' This would decrease stability and actually slow down the aircraft! You cannot just assume that a beautiful flush inlet three inches from the top of the wing will provide adequate cooling. My measurements during development of an oil cooler system showed terrible pressure recovery during low speed.

"I should point out that because with a canard aircraft both surfaces are lifting wings (the canard actually has a much greater wing loading than the main wing); their size, position, interference with each other, high lift devices, etc., have a very important effect on the CG range, the flying qualities and low speed performance. Their design is far more critical than with a conventional aircraft with one main lifting wing (sized for performance, etc.) and a tail sized merely to provide adequate static margin and sufficient CG range. For example, a formula-one racer has an extremely small tail—but it can be designed for one CG only and still provide adequate stability and sufficient control. But if it were a canard, the designer would have much less room for change, to provide a large flight envelope (speed, range and maneuverability) even for one CG.

"Therefore, I am unable, without conducting the appropriate tests, to answer a question like 'Is it okay to move the canard down eight inches to clear my extra radios in the instrument panel.' I am not averse to you making the change, however, if you are willing to conduct the tests and verify satisfactory results. The car-top wind tunnel system is an excellent method; others are also valid.

"Remember, this aircraft was not developed by guesswork, but by a very careful design-test program. Small changes can be full of surprises. If you modify an aircraft, when it is ready to fly you are an *experimental* test pilot, not a *production* test pilot—be prepared to accept the full responsibility to safely plan and conduct exploratory testing and critical flight envelope expansion, for there are no proven limits on your airplane.

"I don't mean to inhibit progress, only to promote valid development. In this way we are also promoting education, which is what EAA is all about!"

FIRST OF THE VIGGEN BUILDERS

Mike Melvill (Fig. 3-9) and his wife Sally, both originally from Johannesburg, South Africa, have been an integral part of canard development since they first saw Burt Rutan's flight demonstration of the prototype VariViggen at Oshkosh. Mike was the first builder to complete his VariViggen and at the time this book was prepared, he had logged over 410 hours on N27MS (Rutan's Design #27 and Mike/Sally for MS).

Before Mike Melvill purchased his VariViggen plans, he had joined the EAA and had a Cougar ⅔ built in his sitting room in Anderson, Indiana. The wings were in his bedroom. This was not acceptable, so he sold the parts and purchased a Cougar 90 percent finished, put it in the air and flew it for four years before he saw Rutan's Viggen.

Melvill (and it is spelled without an "e;" Mike says that his ancestors in Scotland had a violent argument four generations ago, dropped out of the clan, and also dropped the "e" from the family name) saw Rutan's Viggen demonstration and said, "That's the airplane for me. It's not as fast or as fuel efficient as some, but it'll carry two people with three suitcases in comfort."

To accommodate the new project, the Melvills purchased another house in Anderson—one with a 18′ × 32′ family room that was used for building. How long did it take? Mike rattles off the answer with a grin: "Three years, one month, twenty-two days. I don't know how many hours of work, but it was somewhere between three and four thousand. My out-of-pocket cost was $13,500 (in 1977 dollars), including a zero-time engine."

Mike learned to fly in Indiana when the tool and die shop where he was working needed a salesman who could fly to visit customers. During his Viggen project, he went directly by Rutan's plans until he got around to the landing gear retraction system. After breaking retraction cables during construction, he redesigned the entire system and built the complex parts on weekends at the tool and die shop where he was the foreman.

"The original design didn't work out for me and the one I designed was damn difficult for the average guy to build. I had the tools and the experience in the machine shop. I did all my own welding and machining. I made every part of the Viggen literally, except the fiberglass nosebowl and engine cowling. It was a tremendous education—better than college, I think.

"The Viggen has really been a part of our lives, Sally's and mine. We've flown it to Key West and Seattle; we've been in Oshkosh twice, and now we're both working for Burt in the California desert at Mojave with the Viggen, our No. 1 means of long-range transport (Fig. 3-10)."

While the Melvills were constructing their No. 1 Viggen (Fig. 3-11) in Indiana, Rutan was a visitor. He spent half a day crawling around looking at the project. Later Mike flew the airplane—the first one that Burt had ever seen fly except his own. Later they talked at Oshkosh and Mike was invited to come and work for Rutan. When he found out that Sally was a bookkeeper and the couple had always worked together for the same company, he said, "I need *her* worse than I do *you*."

So the Melvills moved west, VariViggen and all, and purchased a home in Tehachapi, 17 air-miles west and 1200 feet higher than Mojave where you don't need air conditioning, even in the

Fig. 3-9. Mike Melvill, left, assists Burt Rutan during a Long-EZ checkout. Melvill was the first builder to complete a Viggen and, together with his wife Sally, has moved to the high desert of Southern California.

summer. The couple commute to work in a rare two-cylinder, 60-hp Aeronca 7ACA "Cheap Champ." The flight takes 15 minutes downhill and 20 minutes on the return. When they are forced to drive, it's 30 minutes each way (26 miles on a winding road) and nowhere near the fun!

Sally Melvill (Fig. 3-12) has a number of "firsts" in aviation. She's the first woman to solo the Viggen, and she has 40-50 hours in that ship as of this writing. She was the first woman to solo the Long-EZ and does her economy practicing solo in the "Cheap Champ." Sally likes this little taildragger because it is sort of a challenge to fly, and Mike likes it for the miniscule fuel bills and because it is tandem (like his Viggen) and would be easy to land in the desert. Sally also has the glider rating in the family. "When I want to fly a glider, I go with her," grinned Mike.

Mike's mother Isobel visited the Mojave area recently and took her first lightplane ride ever with her son in his Viggen (Fig. 3-13). Later she went along to tour most of Southern California, Arizona and Nevada in the back seat of the Viggen. Mike's father, who died before his son ever became involved with aviation, was in the South African Air Force in WW II flying Mosquitos on photo missions.

"I wish that my Dad had been able to see all this," said Mike.

The details of Mike Melvill's VariViggen construction, and particularly his contributions to the landing gear design, were well documented in early editions of the *VariViggen News* and in subsequent issues when the name of the quarterly publication changed to *Canard Pusher*.

Fig. 3-10. Mike Melvill pulls in close to the photo plane in the VariViggen as Dick Rutan follows with the Long-EZ.

In early newsletters, Burt Rutan described his own three gear-up landings in the prototype N27VV in this manner:

"World's first VariViggen gear-up landing. It occurred during the airshow demonstrations on Saturday and Sunday for the fly-in. On Sunday, I had completed the airshow demo, all except the landing. When I moved the gear handle down, I heard a different noise and noted that although I had electrical power to the main gear (transit light on), the mains did not come down. The failure was later determined to be the spring that connects to the uplock arm.

"The spring had apparently been knicked with pliers when forming the hook on one end in 1970 when the spring was installed. Five years later, during the airshow demo, the spring broke. Without this spring, the right main gear remained locked up. After several passes over the crowd, for inspection of gear position and some radio discussion with those on the ground on whether or not to land in an adjacent grass field, I decided to land on the main hard surface runway with the nose gear down. This was taking the risk that the nose gear would not fail and thus reduce the damage. I make a 'full-stall'-type landing with engine and switches off, and after a short roll/slide, I got out to inspect the damage. Damage

Fig. 3-11. Mike Melvill flies formation with Dick Rutan, VariViggen and Long-EZ, over the desert near Mojave. Dick, a former USAF fighter pilot, taught Melvill how to fly a tight formation.

72

was limited to one skid (VS1 extension with small wheel), a small scrape on one wing tip (only one rivet damaged), and partial collapse of my centerline fuel tank. The tank remained attached firmly on its mount and did not leak. The nose gear took the 'slap down' load well with no damage. About 20 volunteers lifted the aircraft up while I scampered underneath to manually free the uplock and to lock the gear down. I then taxied back to a hangar, inspected the aircraft, elected to pin the main gear down and locked for the flight home. Within 1 ½ hours of the gear-up landing, I took off and flew it back to Mojave where I was greeted by 60-knot surface winds. Landing and taxi-in were uneventful despite the fact that at the time two other aircraft were being jerked from their tiedowns and suffered wind damage much greater than my earlier gear-up landing!

"I learned a bit from this experience:

"1. Inspect uplocks during preflight and use appropriate quality control during their installation.

"2. If faced with a main gear-up landing, pull the main gear circuit breaker, extend the nose gear and make a landing with the nose quite high (full flare) on a hard surface. This landing is really not more difficult than a conventional landing and you can expect very little damage.

Fig. 3-12. Sally Melvill handles work in the Rutan Aircraft Factory office when she isn't out flying one of Burt's designs or the family 60-hp Citabria.

"3. Gear-up landings on VariViggens are far safer than on conventional aircraft where one of the first things to get damaged is the carb and fuel line and the possibility of a fire exists.

"4. Note that the emergency extension free-fall system backs up an electrical failure and mechanical failure of the electrical motor and gear box, but does not extend the gear with a jammed uplock. I do not recommend a design change of any type since the gear has had nearly 1000 satisfactory cycles during all types of weather and flight conditions. Any change now would be starting at zero experience with a resulting increase in risk.

"5. Gear-up landings have a very positive appeal from a marketing standpoint. It emphasizes how rugged the structure is to survive with only minor damage. We immediately received seven orders for plans from people who saw the landing!

"VariViggen Gear-up Landing Sagas #2 & #3. We again were faced with having to land on our way to Oshkosh this year with the main gear retracted. We landed N27VV on a hard surface runway with the nose gear extended. We got the prop stopped before touchdown and slid out on the rear skids and nose gear. Again, the nose gear took the load with no damage; all damage was limited to the two aft skids and a non-critical scrape on the aileron control arms. Inspection revealed the problem to be the same uplock spring which caused the gear to remain locked up three months earlier at the Corona flyin. We were on our way that afternoon again after pinning the main gear down, and we flew the remainder of the trip to Oshkosh with the main gear down. The next day at Oshkosh we repaired the skids and rerigged the main gear to put it back in operation.

Fig. 3-13. Mike Melvill's mother, Isabel, took her first ride in any light plane on this hop with her son in the Viggen. Later she toured the Southwest in the back seat of this plane before returning to her home in South Africa.

"This time the uplock spring's loop had somehow slipped out of the bracket rather than failing like it had at Corona. The spring loop was returned to the bracket, this time twisting it backwards so its own torsion wouldn't tend to remove it.

"I feel quite embarrassed by having this spring fail—twice! After all, a spring is something to trust—like gravity. I am recommending that you install a simple addition which consists of adding a branch to the existing emergency extension cable. The present emergency extension cable removes the electric motor from the system, allowing the uplock return springs to push the gear overcenter so it can freefall down. This only backs up a failure of the electrical motor and cannot extend the gear if the uplock springs fail or if the uplock would jam. By simply adding cables to the existing emergency cable and routing them to the top of the uplock belcranks, the emergency handle would not only remove the motor, but would pull the uplocks out and force the gear overcenter and on its way down. Thus, the emergency handle overrides a spring failure *and* any jam of the gear.

"Gear-up landing #3 is of no real concern to builders, since it does not involve a problem which can occur with your aircraft, since an obvious design improvement was incorporated into the plans before they were first released. The failure allowed the MG5 bolt to slip past the MG29 bolt during gear retraction. As such, the microswitch on MG29 was not activated and the gear motor continued to run, jamming the gear way over center and failing the cable. This failure occurred on the third flight of the day at the EAA Western Flyin at Tulare, California, with Bob Eldridge in the back seat. I had taken off to compete in the spot landing contest. Since there was a lot of activity on the runway at Tulare, we decided to fly to another airport about 20 miles away to do our gear-up landing there. The landing on the nose gear and aft skids was uneventful (routine?); the gear was fixed, and we flew back to Tulare to compete in the spot landing contest.

"Now—I don't expect to hear from any more of the VariEze fans about wanting to retract the main gear!"

CHECKOUT BY THE DESIGNER

Since Melvills' VariViggen was the first to fly, its progress was well documented in early issues of Rutan's quarterly *VariViggen News*. Taken in chronological sequence, some of these reports show how the new airplane evolved.

Prior to attempting his first flight, Mike prudently visited Rutan in California and flew with the designer in the prototype. These were Mike's comments as published in the newsletter.

"Burt made the first takeoff and one of the lasting impressions was the sight of the shadow following along in the early morning sun.

"Burt demonstrated level flight, slow flight, turns, steep turns, and most important, pitch trim changes with abrupt power changes. This is something that has been emphasized over and over and rightly so. It is an unusual condition, but to be perfectly honest, not a difficult thing to get used to. Personally, I had very little problem with it; however, I was thoroughly aware of the condition and I am very current in several different aircraft. This is a point that cannot be too strongly emphasized.

"Any person thoroughly checked out and confident in say a Cessna 182, a Grumman Tiger and a tail dragger, in my case a homebuilt Nesmith Cougar, will have no problem with the VariViggen.

"After a little stick time in the back seat, we traded seats and I spent quite a while just taxiing the airplane all over the place and let me say this, there cannot be a more simple or manageable airplane anywhere. It is so easy to drive around on the ground and it goes right where you point it. Marvelous!

"Then I did some high speed taxi runs; again just point it and go; no problem with keeping it on the centerline as it tracks perfectly straight, and the rudder becomes effective very early in the take-off run.

"Next we tried some nose wheel liftoffs. This must be done in accordance with Burt's instructions in the owners manual. Get it stabilized at the speed you want, retard the throttle, then rotate. The nose will come up and is very easy to control. I want to emphasize, pitch control is excellent. Before I tried it, I was worried that pitch control may be marginal. However, this is not so at all. Pitch control is really great; you can put the nose anywhere you want to and maintain it there.

"Then we did a couple of runway flights, liftoffs and flying in ground effect. Again, control is excellent, both pitch and roll, and I felt very happy in it. Full power takeoff was an anticlimax—it was very normal and flew just like any other high performance single engine. Handling qualities in the air are great. It flies perfectly in my opinion. In fact, I was very pleasantly surprised. It is all I had ever hoped for and more.

"The landing, again was almost anticlimactic; with the correct airspeed and altitude, it will land itself. The only thing to remember when landing is *not* to try to full stall land it as you would a Cessna. It is much better to fly it on; the gear is very forgiving and takes care of most bumps. Don't try to hold the nose gear off right down to a virtual stop because it will stay up until the canard quits flying and then will fall through rather abruptly. This is no problem, but I personally think that you get a nicer landing by letting the nose down before the canard quits. Also, this gives you better braking, as all the weight will be on the wheels instead of some of the weight being carried on the wings, which it would at the high angles of attack possible by holding the nose off.

"If you try to stall it on, it is possible to hit the tail skids on the runway, so until you get really familiar with the airplane, listen to Burt and fly it on!

"To recap: make sure you read and fully understand the owners manual on test flights. Then go out and enjoy your Viggen, it is a super airplane.

"*Burt's comments:* Mike is a very proficient pilot. He handled the airplane in the first few seconds like he had 100 hours in it. I particularly noticed how well he flew the rudders—must be his Cougar experience. Mike should feel right at home and confident on his first flight in his Viggen."

The very next issue had this enthusiastic reply from Mike.

"Well, we finally got there! At 11:30 am, 9-22-77, I took off from the Anderson Municipal Airport, and everything behaved as it should. It stayed up for about 45 minutes, did *not* retract the gear and made a perfect landing. I cannot describe the feeling, it was absolutely fantastic. Thank you so much for a fabulous flying machine! Later the same day I climbed to 5,500 and retracted the gear and checked it out generally.

"As of today, I have 13.2 hours on it with no problems. At 7,500′ in level flight, she trues out at 170 mph at 2700 rpm. I have a 70 × 70 'Ted' prop, but for an 0-360 even that is not enough, as it will over-rev at low altitude. At 3500 feet she will indicate 165 mph at 2700 rpm, but this is not full throttle. Initial climb solo is 1500 f/min. At 5000′ solo she makes a steady 1000 ft/min. All systems operate perfectly, reflex, electric trim, and gear are really first class and I am very satisfied.

"My radios (TERRA 360 com & 200 NAV) are really outstanding, and the tower at Anderson says I have the best transmission of any radio in the area. I cannot say enough about the airplane. She really is a hell of a fine craft. I love it.

"Empty weight, 1252 lbs

"Empty cg 132.78

"Main tank holds 24.7 gals

"Wings hold 6.5 each - 13 gals total

"So far I have flown it at 125"cg, 124, 123, & 122. It handles well in all planes so far.

"Yesterday I loaded 170 lbs in the passenger seat and could hardly tell any difference.

"It takes 16 minutes to transfer 13 gals from the wings to the mains. Cyl head temp runs between 375 & 425, oil temp 165°, ground handling is excellent, brakes are very good. Rotation with full throttle occurs with full aft stick at 70 mph indicated. Initial climb at 85 mph indicated for gear retraction, then trimmed down to 120 mph for good cooling gives 100 ft/min. R.O.C. Canard stalls at 60 mph indicated (airspeed may not be accurate at slow speed) will climb with canard stalling & unstalling, and is fully controllable. Side slips well.

"Actually, it flies just about like yours. I must say I really get a heck of a kick out of flying it. I will enclose some pictures. I painted it off-white with dark green trim.

"Sally sends her best regards. Thank you again, Burt, for making it possible."

Three issues later (nine months), Rutan reported that Mike and Sally have really been giving their Viggen a workout.

"They visited RAF in May during a 600-mile trip. Mike and Sally were alternating front/back seat pilot chores. They were loaded down with baggage and handled a 9000-foot density altitude takeoff at Albuquerque with no problem. His summary after arriving home: 'Once again the Viggen has proven to us what a really practical cross-country machine it is. We love it and would not trade it for anything.'

"Mike designed and built a very clever angle of attack instrument for his Viggen. Instead of a potentiometer on the vane (the one that's so hard to find), he made up a wiper with three electrical contacts. These go to three lights on the visor arranged in a vertical format. When the center light is on (green), the airplane is 'on speed.' When the top light is on, you are too slow; too fast if the bottom one is lit. This system automatically makes you fly the correct approach speed regardless of weight. It works exactly like the indicator lights in an F-4 jet fighter, yet Mike built it for $5.00!

"We have been able to inspect the Melvill drawings for the worm-drive main gear modification. They do add some complexity, but in the long run we feel that it's well worth it. We highly recommend it and plan to incorporate it when the Viggen plans are updated in the 2nd edition."

Four more issues went to press and then Mike penned the following report for VariViggen and VariEze builders:

"I flew N27MS to Oshkosh this year and had a super trip, flew in formation all the way there and back with a 180-hp Grumman Tiger, piloted by Sally. The Viggen had to be flown at quite a low power setting in order to stay with the Tiger at ground speeds around 140 knots (162 mph). I only burned 7.8 gph average for the whole trip. Not bad for 180 hp.

"We flew from Mojave via Las Vegas, NV, Provo, UT, Scottsbluff, NB, Rochester, MN, to Oshkosh. The Viggen joined up with the Defiant and Long EZ for several airshows during the week. From Oshkosh Sally and I flew (Tiger and Viggen) to Indiana to visit family and then via Coffeyville, KS, Tucumcari, NM, Abq, NM, to Mojave. It was a most enjoyable trip. I put 37 hours on the Viggen; she now has 366 hours and apart from adding two quarts of oil, she required no maintenance. The only new Viggen flying since the last newsletter is a French VariViggen, built as a flying test bed for the Microturbo jet engine. This very beautiful aircraft is powered by two of the diminutive jet engines (same as BD5-jet) located one above the other. The aircraft has only flown a few times, but reportedly is quite fast.

"Unfortunately, since the last newsletter there have been two VariViggen acidents. Although causes are not known for sure, pilot proficiency still appears to be a problem. You must be current and *sharp* in several airplanes before attempting a first test flight in any new airplane. *Do* follow the owners manual to the letter. *Do not* omit the high speed taxi and runway flights. If possible, get a checkout in a Viggen with an experienced Viggen pilot. All of you are spending several years and several thousands of dollars; don't throw it all away with careless flight testing.

"I have a hunch that several Viggens are almost flight ready. When you are, give us a call; we will be glad to help you with your test program or provide a Viggen checkout."

In the next newsletter, Mike had this to report about Sally:

"N27MS has flown regularly and on June 19, Sally soloed our

Viggen for the first time. She had flown it regularly from the front seat, but I have never had the guts to get out and let her go solo! I finally could not put if off any longer, and she went out and made three perfect landings. Sally's total flying time is 120 hours, mostly in C-150's and with a little Grumman Tiger time. The only problem is now our Viggen will not always be available for me to fly! Congratulations, Sally."

BUILDING THE VIGGEN—NOT VERY EASY

George Craig is a retired school psychologist living in Milpitas, California (Fig. 3-14). He is one of the VariViggen builders whose airplane was perhaps ¾ completed at the time we contacted him (Fig. 3-15). Craig has been a pilot for many years, working on his own Cessna 172, including rebuilding the engine. He installed the popular 180-hp Avcon engine conversion kit on his Cessna, doing the majority of the work on it himself.

We asked George to share his reasons for choosing the Viggen for his first homebuilt project and any other thoughts or suggestions he might have. Here's what he had to say.

"As to why I selected the Viggen—that goes back to when I was in high school and was doing designs of airplanes before WW II. I had always wondered why one puts the tail behind the airplane. If you look at an airplane, it's like an arrow in reverse. The wing should be the stabilizing part of the airplane and instead you put the tail in back, and you end up with a negative lift component. In other words, you're always dragging that tail along behind you and, in

Fig. 3-14. George Craig checks a landing gear fitting for his homebuilt project.

order for it to control the airplane, it has to work in reverse so that you actually lose lift. That doesn't really seem reasonable to me and the Wright Brothers must have felt the same way because they certainly put the tail out in front.

"In selecting the Viggen, I wanted an unusual design. I had heard that the VariViggen would go very slow and make quite steep turns without the possibility of stalling. And I found out later, in the two times I flew Burt Rutan's airplane, that this was the case. So, on the basis that it was an unusual design, one which would theoretically be very efficient and the fact that I had an interest in the canard concept dating back to high school, I selected it. Another factor is I felt comfortable with woodwork as that is what I was most proficient with and had the most tools for. So there again, since the fuselage of the VariViggen is spruce and plywood, it appealed to me.

"Another point: at the time that I started the Viggen, there wasn't the complication of having the choice of the VariEze— though for reasons I'll indicate later, I wouldn't have chosen it anyway.

"It was very obvious to me from the beginning that the cost was going to be in the thousands of dollars. There was no possibility of this being built with any degree of efficiency and not

Fig. 3-15. George and Madge Craig work together on this homebuilt design. Without family cooperation like this, many homebuilt projects never fly.

buying all the possible components prefabricated that you could. As you are probably aware, in the homebuilder movement the majority of the homebuilders are 'scroungers,' and as a result they take great delight in taking something and putting hours on it and modifying it and making do with it. I still feel that I'm essentially first a pilot and then a mechanic. As a result, I want to get the airplane in the air—I want to see it fly. And I've taken all the shortcuts I can, though they're quite expensive. Offhand, I'd say that probably the Viggen will turn out to be one of the most expensive homebuilts that a person could approach.

"Here is some idea of the cost of the Viggen. On mine, I find that the instrumentation and the wiring on the airplane alone (unless you try to scrounge the parts) will cost you very close to $2,000. The engine—in my case I've always rebuilt a first runout engine—will cost me $2400 to obtain and probably another $1,000 in parts. That's very conservative. I think by the time I have the alternator, generator and the whole thing, I'll be well into $4,000 for the engine. And that, believe me, is cheap because these engines are selling up between six and eight thousand dollars.

"Engine availability depends on what you put in them. The Viggen as it's originally designed was intended to use the 150-hp Lycoming or the current 160. Now those engines are a dime a dozen; you can pick them up almost anywhere because that's the engine that is replaced when you go to the Avcon conversion. I found that the 0-360 180-hp Lycoming makes it possible to use original wing plan—the swept wing—rather than the later high-performance wing. This retains the airplane's flying characteristics; it retains the better stall characteristics of the airplane, and by adding the 180 horsepower, you get performance so that you can fly in and out of high mountain summer strips.

"An additional, but certainly very individual reason for my choosing this airplane is perhaps a little bit curious also. You may have been around enough fly-ins to run across airplanes like the ones labeled "nostalgia," and you find airplanes that while they are not an exact replica of war planes of the pre-WW II era (the bi-plane era), they are obviously intended to be very similar. You'll see planes painted up in Army colors and attempts to have something like those early planes. You'll see a J-3 Cub painted to look like it was being used as an observation plane for D Day—that sort of thing. Well, for me, I haven't seen any airplane that could mimic with any degree of reliability a modern military jet fighter. But if you take a look at the VariViggen, with the exception of the canard, you'll find that that airplane very much resembles the F 14 Tomcat fighter and the later F-16 Air Force fighter.

"To my way of thinking, it has the possibility of being dressed up in Navy colors and Navy stencils for emergency exits, etc. What I'm saying is that for me it is a further step along the line of imitating military airplanes, which has been part of the movement for a long time. Now, I admit that it's certainly taking on the nature of a toy. But that's what you're in the homebuilt movement for. I certainly don't want to be in this for work. I want to get all of the fun I can out of it, and part of the fun for me would be to fly it into Moffett Field (U. S. Navy airfield in California) on Navy Day sometime and have some of those young military pilots come over and gawk at it and 'oh and ah!' Well, that's my specific reasons for wanting the Viggen. Because so far as I know, there's no other airplane that really resembles a jet. There just isn't any. None of them have the wing far enough back, and you don't sit far enough forward in splendid isolation, you know, as you do in the VariViggen.

"I've been over to Mojave to see Rutan and I've always been treated very nicely over there; I appreciate what they've done in back ups and support, although the Viggen, it seems to me, is losing some of that support. It's probably as much as anything the fault of the VariViggen builders. They do not communicate their progress to Rutan. There are dozens of VariViggens around the country that are in the process of being built. But all of us, I think, are a little ashamed of how long it takes us. This is a fearfully complicated airplane, and it is going to take you time and time and time (Fig. 3-16). A good example will be my experience with the retracting mechanisms (Fig. 3-17). I was very clever in setting my entire retracting mechanism up on the main spar and making sure that it worked well before I installed the main spar in the airplane. And, if I do say so myself, I had a beautiful installation which was

Fig. 3-16. George Craig uses all of his two-car garage in Milpitas, California, to build his VariViggen. Power tools for the shop were purchased to go with this project.

exactly according to the plans. Well, I found that there were some aspects of the plans that in small details were causing problems— such as the fact that the cables were rubbing and an additional pulley would have had to have been installed. But worst of all, I discovered that the entire gear system had problems. I was playing around with the apparatus one day and let the electric motor that potentially raises the gear get a little bit beyond where it should have and snapped the cable. They are very tiny cables. Then I began to hear that other people were having trouble with these cables and the only way that you could get the gear down was to pull the handle, drop the gear, and hope it would go down and lock. I didn't like that idea. And there were more problems of this kind.

"So what happens? Here's this complicated mechanism which has taken me months to build and the whole thing, or 80 percent of it, has to be junked. I started over from scratch again to build a different mechanism which was developed by Mike Melvill and

Fig. 3-17. Landing gear fittings lie on the bench as George Craig checks alignment of this canard. EAA posters and aviation photos cover the wall of his fold-out garage door.

involves worm gears and little transmission cases and this sort of thing. This is again fearfully complicated and required you to have several friends who are machinists or you're going to be in trouble.

"I feel one of the problems is that I don't know how many parts of the VariViggen were simply designed by Rutan but had never been tried out. If you take a look at the original VariViggen, you'll find that it is built differently than the plans. Mike Melvill, who is the only one that I know whose Viggen is flying (although I understand that there are others) had to depart from the plans. There's a friend of mine now down in Texas, Leonard Dobson, who completely revised the retracting mechanism. He did this on the basis of a complicated transmission case and he uses a bicycle chain. There are a lot of airplanes that use that on the retracting mechanism. Leonard uses that on his main gear.

"I think that some of us have been ticked off and discouraged by the fact that we've had to go back to square one on the most difficult and complicated part of the airplane. However, mine is now well on its way; I've got everything out to be machined and I think we're in pretty good shape. On the whole, I know very well and I feel quite confident that if I have any problems, I can go directly to Rutan and get a good answer and a good fix for the airplane. I feel confident of this, and as long as he's in the airplane business, I'm sure he will provide generously that kind of advice and consultation. In addition, Mike Melvill, who's in charge of the VariViggen project in Mojave, is equally an excellent person to consult. He is a very practical person and I have enjoyed my contacts with him.

"I was president of Chapter 6 of the EAA in San Jose last year, and I've maintained close contact with that group. However, your friends, advisors and the people that can help you with your homebuilt are in a constant state of flux and change. Right now the person who is by far the most important source of assistance to me on the increasingly more sophisticated and technical problems would be Ray Stevens here in Santa Clara who is a master airplane builder. To my knowledge he's built a Marquart Charger, a Hyper-Bipe, and I'm helping him now in building a Christen Eagle. He's also working on some other projects. He has a fine shop and is very helpful and generous with his time also.

"I'm not going to use the canopy as it's in the plans. Using Mike Melvill's new type of retraction mechanisms, I'm making the rear cockpit so that it will have brakes and, as Mike says, it will be possible to fly the plane to a full stop from the back seat, which is not the case with the plans. I've changed the seating and the interior, which I think everyone who works on the Viggen has. I'm much more elaborate in instrumentation because I want to be very sure I know what's happening back in the engine compartment. However, I've stuck as closely as possible to the original plans, and it's been possible to stick fairly close. Like, for instance, the plans have a hinged nose cone (Fig. 3-18)—I plan on attaching mine with camlocks. That sort of thing— nothing major.

"As to how much time and money I've spent on the project to date. I have spent parttime for a period of four years, off and on (Fig. 3-19). I've spent the last six months purely on the VariViggen. But I've found something important to me personally. I've found that what I'm missing as a person who comes from a professional area (social service profession, for example) into working in a mechanical field is the shop techniques. For instance, what do you do when a hole is drilled off-center and you need to redrill a hole; or you find your countersink doesn't center on the

Fig. 3-18. Although VariViggen is designed with hinged nose, some builders choose to omit this feature and substitute camlocks.

hole and starts to drift off because it isn't a piloted countersink? Well, you find out that people who are shop 'pros' know that you just take a thicker piece of steel, bore a hole in it and use that as a drill guide for countersinking with no problem. Well, it's just hundreds of little things like that that I'm learning because for the last two months I've been working everyday for six or seven hours helping to assemble a Christen Eagle for one of the ladies in our area here. Ray Stevens is doing the work and I'm working as a full-time assistant—thus learning many, many shop techniques which when I'm ready to come back and work on the VariViggen will help me. So you see what's happened. First I worked on it part time, full time in the summer, and then I worked on it full time since my retirement last June (except for vacations that I've taken), and here I am four years along the way, and I expect it will take another two years to do this.

"I think that Rutan's designs are excellent. I really feel that they are in the process of development, and anyone who gets the idea that he can take a Rutan design when it first comes out and follow that design and get through the projects without making changes and without some frustrations because of design difficulties, is out of his head. It's going to have to be a process of modification. I don't see that as a weakness. I think that's true of any aircraft development. You simply have to continue to make changes.

"And as to hard-to-get parts—for the most part, they're available. The machining that's required for the new Melvill-type landing gear is extensive and if you do not have available a machine shop with at least an end mill, you'd be out of luck. You just couldn't do it with hand tools." (*This gear is now available from Ken Brock.—Ed.*)

"As far as flying the Viggen is concerned, there have been too many people who have had trouble with the Viggen due to the fact that the engine produces a high thrust line which causes the airplane to act in reverse. You increase the power and the nose goes down; you take off the power and the nose comes up. I think the plane is moderately sensitive on the controls; I think it is just about right. The controls are much better balanced than they are, for example, on the Hyper-Bipe which is an acrobatic airplane and will drive you wacky trying to control it. But to take on the whole, it's this peculiarity that gives trouble. And unless you're used to flying Lake amphibians or some other high thrust line airplane, I think it would behoove you to take the precaution of going over and flying with Mike Melvill or getting one of the Rutan factory pilots to come over and check you out. I think you need to fly that airplane from the front seat with somebody in the back seat who has previously had VariViggen experience because I'm quite positive that it is no more difficult to fly than the original Cardinal which had the problem with over-sensitive longitudinal stability. You simply need to have just a little time to get your muscles in tune with the idea that when you pull that throttle off, you by gosh better shove that stick forward. It becomes a reflex after you're used to it with a little bit of training. As far as the landing in this airplane is concerned, I think it must easily be the easiest airplane of the homebuilts to land that you could possibly get. There's nothing to do with it except close the throttle, hold the stick back in your lap, use the throttle to increase or decrease your glide angle, and you simply bring it down and land it. There is no stalling of this airplane so that the landings become quite simple.

"I do not think that this design is too much airplane for many builders. It's simply that if the person is a 172 driver, he's gotten used to the idea that when you pull on full flaps and apply full power, the nose just comes the hell up and you've just about got to push that wheel through the instrument panel to get that nose down. Now, we don't consider that as being a difficult airplane to fly. You

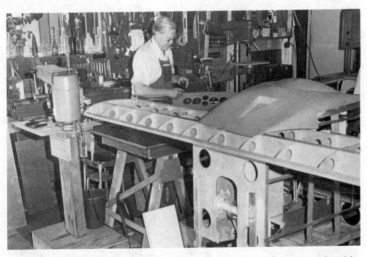

Fig. 3-19. Canard and fuselage near completion in George Craig's comfortable, well-equipped shop/garage. Note orderly arrangement of tools on the wall in the background.

can check with other people and verify this, but I think you'll find that the VariViggen is no worse than a 172 with full flaps. But believe you me, all the person needs to do is get over that one psycho-motor learning experience, which is no more difficult than a 172 full-flap-go-around and you're in business. So you see, it's not a case of there being too much airplane. The airplane when it's properly handled and under normal flight conditions is practically stall-proof. You can't get into the problems you get into with most

Fig. 3-20. Every available space is being used for instrumentation and nav/com equipment. George Craig is spending $2,000 just for instruments to monitor the engine compartment and provide for occasional IFR.

airplanes, and it's an easy airplane to land. What does it leave you with—just that high thrust line problem.

"You ask whether I could go into the Long EZ and/or VariEze design. This, incidentally, was not on the market when I picked up on the VariViggen. My answer is a definite and resounding NO! First, let's take the VariEze. As I pointed out, it's costing me about $2,000 to adequately equip my airplane with sufficient instruments to monitor the engine compartment and to provide for occasional IFR (Fig. 3-20). There's no way that I could afford to properly equip the VariEze, even though I am planning to spend approximately $15,000 to $18,000 on this project. For example, an electric artificial horizon will cost two to three times as much as the war surplus type. Besides that, the instruments on the VariEze (the Long-EZ I'm sure is in the same boat) have got to be miniaturized. They've got to be the very most expensive instruments that you can get for that airplane. The VariEze, itself, is absolutely out of the question for a person like myself who weighs 200 pounds. That airplane simply will not carry me with any amount of baggage; it's absolutely crowded. You have to use special luggage—the whole bit. Now, it's a wonderfully efficient airplane for someone who wants to stay on smooth pavement. Well, I don't necessarily always want to fly that way. The VariViggen with its 550 main gear can, according to what I've been told, be used on a fairly rough field—nothing excessive—but it certainly is a lot more adequate than the little tiny wheels that you'll find on the VariEze. I presume the Long EZ will use a comparable landing gear.

"I imagine the VariViggen can possibly be operated for about the cost of a 172; it's a little bit more efficient than the Avcon conversion. It's about 25 miles faster. I think that by cruising the VariViggen back to about 8½ gallons an hour for cross-countries, it will do quite well. With our gasoline situation, I'm hopeful that gasohol for aircraft engines might be our solution in the future. It would seem to me that the most one might have to do would be some minor modifications to the carburetor. Besides, most homebuilts are not flown a great deal—they're mostly for hobby and sport flying to fly-ins, and spend an awful lot of time in the hangar.

"As far as help from other people, I get a great deal of help from my brother-in-law, Paul Schillerstrom, who is a machinist, retired, and from Ray Stevens whom I've mentioned. They both have helped me a great deal. And when I need another set of hands, my wife has been extremely generous. Madge comes out in the

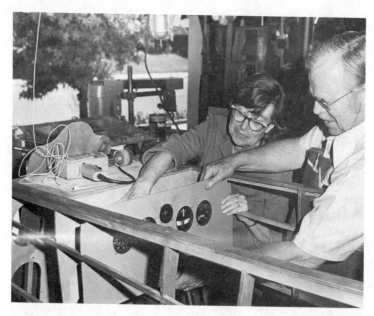

Fig. 3-21. George and his wife, Madge, check out the front panel in their VariViggen. The couple owns a Cessna 172 that they use for extensive cross-country flying.

shop and helps (Fig. 3-21). She has been very supportive. The EAA has been helpful as well. John Winters who was our EAA designee before he started to work on United night time, would come around periodically and make inspections. On the whole, I think the EAA Chapter is more socially oriented. I don't get an awful lot of direct help, and it is not necessarily reliable because many of the people in the organization are floundering as much as I am.

"You ask me if I would build another aircraft after I completed the VariViggen. I most certainly would. Now you know the VariViggen allows you to do all the mechanical and instrument and radio and wood work and the composite construction, but it doesn't really give you the opportunity to work with fabric. I have sitting down in my shop right now a Smith Mini-plane Special, a single-seat bi-plane, which is just sitting there waiting. Most of what needs to be done to that plane is just engine installation and fabric covering, which will be an interesting project when I get finished with this. Then I hope also, if time allows and health holds out and all that sort of thing, that we can restore an antique. So I think that you can say that for us having some sort of an aircraft project in the shop is going to be necessary."

Chapter 4
VariEze—It's Really Very Easy

Building and flying a VariEze is certainly a new way of life. Nat Puffer of Minneapolis penned the following account of the airplane that changed his life. We reproduce it, incomplete sentences and all. Then we'll look at where this design came from.

"This is a special 'thank you' letter. The VariEze has really changed our lives. Much has been written about how easy and satisfying it is to build a VariEze, and much has been written about how delightful it is to fly a VariEze. With all of this we concur. But not enough has been said about how owning and flying a VariEze can change your personal lives.

"When you are building an airplane, you live in relative obscurity, except for other builders. You tend to shun social obligations to get the plane done, and tend to drop out of things going on around you. But once the airplane is completed, especially a VariEze, it is an instant passport to fame. You become an instant celebrity! It is very ego-building and in stark contrast to what has gone before.

"It starts with offers of free use of airplane trailers, free hangaring and/or reduced rates. Very prompt attention from the FAA, etc. Every time we open the hangar doors, an instant crowd of admirers gathers. People come back time and time again, and just stand and look. Special recognition by the control tower. All kinds of people anxious to help in any way possible.

"Being reported as a UFO, and being interviewed on the radio. Being the subject of a sound movie, and being interviewed on the local TV station.

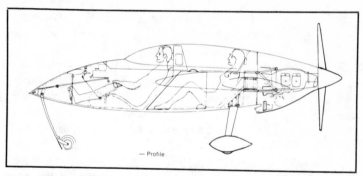

— Profile

Fig. 4-1. Profile drawing of the VariEze (courtesy Rutan Aircraft Factory).

"At Oshkosh, receiving a very warm and friendly reception by the control tower, and being personally escorted to a parking space right in front of the main gate. Being asked to participate in a fly-by. Being asked to participate in photo flights in formation for *Popular Mechanics, National Geographic* and *Sport Aviation*. Being asked to announce on the Oshkosh PA system for the VariEze fly-by just before the airshow.

"On return home from Oshkosh, having Minneapolis Flight Watch carry on a running conversation for 90 miles asking if I could land at Minneapolis International so they could see my VariEze. At Voyageur Village being swarmed with kids. At a local fly-in at Osceola drawing the crowd away from all the other airplanes. Not being allowed to pay for my gasoline. Having the 3M employee magazine editor asking for an interview, and having to fly a special photo flight for the company photographer.

"Having everyone who flies in the other plane on a photo flight ecstatic over how much the VariEze excels any other plane in the sky. (I make gunnery runs and fly circles around the chase planes.) Being very proud to be part of the VariEze program.

"Best of all, how ecstatic my wife is about being a part of it all, and how she simply eats it up! She delights in explaining all the intimate technical details to people, and how wonderful it flies. (She has never even had her hands on the stick and would be terrified if I asked her to fly.)

"There is no indication that any of this will every let up. It has truly been one of the nicest things that has happened to us in our lifetime."

HOW IT ALL BEGAN

Burt Rutan detailed the development of the VariEze in a first-person report published originally in *Sport Aviation*, the popular and colorful monthly magazine of the EAA. The evolution of the VariEze as described by the designer-builder follows, with permission of both author and publisher.

The VariEze (pronouned "Very Easy") design started in early 1974 (Fig. 4-1). At that time it was a high wing/low canard configuration. Its structure was all metal. The prototype was never completed, partly due to complexity and weight problems with the structure and due to spiral instability discovered during model tests.

Further wind-tunnel model tests (Figs. 4-2 and 4-3) conducted in the fall of 1974 led to the general arrangement of the VariEze. This aircraft, N7EZ, was built in four months and made its first flight in May, 1975 (Fig. 4-4). The main aerodynamic benefits sought were the following: the canard and wing systems could be carefully designed to provide natural passive angle-of-attack limiting to make the aircraft departure and spin proof. Inclusion of the new NASA-developed winglets could reduce induced drag and thus provide better climb and cruise efficiency. The layout, using canard elevons for roll and pitch control resulted in an unusually simple flight control system. Packaging of the two occupants was efficient and resulted in a considerably lower wetted-area airplane than a conventional design.

Oddly enough, the decision to use the all-composite glass-foam-glass sandwich structure was mainly an expedient, rather

Fig. 4-2. A recent scale model of the VariEze is "flown" in the NASA wind tunnel. This ¼-scale model produced excellent data for further canard development (courtesy NASA).

than an attempt to develop improved technology. Because of the many unknowns, the prototype (N7EZ) was considered a research airplane to develop the canard aerodynamics, not the prototype for homebuilder plans. In addition, it was decided to attempt several distance and speed records in its weight class. The structure then, was merely an easy-to-build expedient to allow quick construction of the prototype. It was not known if it would be light enough or durable enough for anything beyond the prototype. However, we found during the detail, design and construction of N7EZ and through the many lab tests of the materials that the structure offered some real possibilities. Structural design philosophy was to design to a much greater strength than normal to encompass variations in workmanship and unknowns about long-term degradation (moisture absorption and ultraviolet deterioration). Then, to provide an acceptable structural weight, the structural configuration was closely optimized by using unidirectional spar caps following the maximum thickness contours and by using the skin to absorb bending and torsional loads. The extra beef for the conservatism was put in the skins to resist surface damage. The result was gratifying—a structure with excellent surface durability, a 12-g ultimate load factor and competitive weight. N7EZ with its 62-hp, 139-lb Volkswagen engine had an empty weight on its first flight of 399 lbs. This later grew to 420 lbs after the addition of extra equipment and had a limited electrical system.

Due to doubtful engine reliability, N7EZ never demonstrated its range capability as originally intended for distance records. It did set a world's distance record in August, 1975, of 1638 miles, only slightly over the old record, using 241 lbs of fuel at low altitude

Fig.4-3. Early cartop wind tunnel model of the VariEze was photographed in front of the RAF at Mojave (courtesy Burt Rutan).

without a mixture control. To attain its full capability in the 500 kg weight class would require burning 495 lbs fuel, using a mixture control, an optimum prop, and flying at optimum speed. This would be a 35-hour, 4000-mile flight, requiring flight at heavy weights in darkness. We had found that, even at reduced power settings, the poor reliability of the VW engine made this attempt too risky. N7EZ had, on two occasions, experienced failures requiring immediate landing to avoid total loss of power. Also, maintenance was high. Dick Rutan installed a 60-hp Franklin in N7EZ to again address a distance record attempt. However, vibration and other engine installation difficulties cancelled that plan.

The performance and efficiency of the VW-powered N7EZ was excellent. Its top speed was over 180 mph. It could achieve over 40 mpg at high cruise speed and over 60 mpg at 95 mph. Its span efficiency (e) at 1.15 was greater than anticipated due to the local aft wing upwash (induced by the canard) increasing the benefits of the winglet system. It had to be parked in the nose-down position to provide adequate ground stability without pilot. To minimize the disadvantage of this unusual procedure, a ball-screw device was installed to allow the pilot to raise or lower the nose gear to kneel the aircraft with him aboard.

In the summer of 1975, after flying N7EZ 100 hours, we addressed the feasibility of offering a VariEze-type aircraft to homebuilders. Several problems needed to be solved:

(1) There was no suitable engine available.
(2) The stall speed was too high.
(3) Roll control at low speeds was poor.
(4) The pilot-in-kneeling system was inadequate, requiring too much effort.

Fig. 4-4. Early air-to-air photo of N4EZ with N7EZ (the No. 1 VariEze) in the background near Mojave (courtesy Don Dwiggins).

Problem No. (1) was a difficult one, since the VW's reliability above 60 hp was in question. The Franklin 60 aircraft engine was out of production. The only alternative was the 173-lb A-75 Continental. At that time they seemed to be in adequate supply on the used market. At 173 lbs, the A-75 was too heavy for N7EZ. It was decided to scale up the design to a size correct for a 173-lb engine. This increased the wing area from 59 sq. ft. to 67 sq. ft. Every dimension in the airplane was changed requiring a completely new prototype.

Problems No. (2) and (3) were both due mainly to the poor lift of the GAW-1 airfoil on the canard. This airfoil with plain flap at the low Reynolds number (half million) had a max lift coefficient of less than 1.6. A new canard wing was built for N7EZ using a new airfoil developed at the University of Glasgow. This G U 25-5 (11) 8 section promised a max lift coefficient over 2.6 when used with a slotted flap, a value unheard of at this low Reynolds number. I did not believe problem (3) would be adequately solved, so I designed ailerons to be installed on the rear wing.

Problem No. (4) was attacked by developing several nose-gear retraction devices ranging from electric screwjacks to oscillating mechanisms like a car jack. None of these were satisfactory. The decision was then made to require the pilot to raise the nose, lock the gear down, and then climb in the airplane. With this in mind, the simplest system was selected—a one-piece rod attached to a block running in a guide for up and down locks.

New flight tests were then done with N7EZ to test the modifications. The new canard wing performed excellently, lowering the stall speed nearly 8 mph and considerably improving roll rate at low speeds. We now considered the roll authority to be adequate and thus decided to retain the elevons. The simple control system with roll and pitch on the front wing would be used on the second prototype. We were satisfied with the simplified nose gear retraction system.

The second prototype (N4EZ) (Fig. 4-5) was designed in the winter of 1975, built in four months and made its first flight in March, 1976. Our search for an A-75 Continental engine was not successful, but we found a good deal on a Continental 0-200. We removed its starter but retained its alternator, resulting in an engine weight of 206 lbs, which required a nose ballast of 10 lbs to achieve the proper CG.

The structural redesign for the second airplane involved additional conservatism in allowable stresses and some other weight penalties for items changed to simplify construction. We had anticipated that N4EZ would weigh about 540 lbs with the 0-200 engine with alternator and NAVCOM radio. The "basic" Eze with an A-75 and no electricals would weigh about 500 lbs. To our dismay, N4EZ tipped the scales at 585 empty. We then anticipated that the average homebuilder's VariEze with an A-75 and no electricals would weigh about 535 lbs.

N4EZ completed its 85-hour flight test program including engine and systems development, flying qualities optimization (all weights and CG's), performance determination, dive tests, spin tests, environmental qualifications, etc., in ten weeks.

The tests indicated few modifications. The aircraft was not susceptible to stall/spin. Overall flying qualities were satisfactory.

As we went to press with the plans in July, 1976, the VariEze was a very basic, simple configuration with adequate useful load to allow 700 miles range with two crewmen. Single-place with one hour's fuel, is a true "hot rod" with an initial rate of climb near 2000 fpm and a demonstrated ceiling above 25,000'. However, as the homebuilder's airplanes started hatching from the various basements and garages, we quickly found the "average" homebuilt Eze to be much heavier than anticipated. Whereas we always pitched the Eze to be light, basic and simple, a large percentage of the builders were loading them up with extras. Of the first five to fly, two had full IFR instrumentation/electrical systems (Fig. 4-6), extra gadgets and heavy finishes, resulting in empty weights near 700 lbs. These were, unfortunately, single-place airplanes due to

Fig. 4-5. N4EZ, second of the VariEze prototypes, was built in four months. It was powered with a Continental 0-200 with the starter removed. Note rear wing ailerons.

their low useful load. Despite our non-stop preaching about weight control in our newsletter, many of the Eze's continued to be built overweight. In April, 1978, we conducted a survey of Eze's flying and found that while a few are at the desired weight, the average Eze is 30 to 50 pounds overweight. Twenty-five percent of them were marginal for two-place operation due to more careful weight consideration by the average builder. We later removed the alternator and the nose ballast from N4EZ to lower the empty weight. The heavy 0-200 and 0-235 engines are recommended only in the stripped-down condition without starter or alternator.

Our most serious problem with the VariEze surfaced when the initial homebuilder's airplanes began flying in March, 1977. We found that relatively minor rigging errors in setting the back wing incidence and twist could overpower the elevons' ability to roll the aircraft. Of the first six Eze's to fly, two required full roll control to maintain level flight and had to be assisted with rudder to remain upright! These aircraft flew normally after rigging errors were fixed or corrected with trim tabs. If the pilots had not been proficient, accidents could have occurred. This problem, of course, indicated the need for an immediate major modification to increase roll authority. Roll control on the front wing is only about 25 percent as effective as on the rear wing because aileron deflection on the front wing causes a down wash change that makes the aft wing oppose the roll input. Adding to the problem was the fact that an elevon, when doubling as aileron *and* elevator, is a poor aileron at large elevator deflection (slow speed). Clearly, we had to abandon the elevons and put ailerons on the back wing. The front control surfaces would be used only as elevators, like our previous VariViggen aircraft. Working under the pressure of a major flight safety consideration, we did all the following within two weeks: design and build ailerons and aileron controls for N4EZ; flight test them throughout the envelope, including stall/spin and flutter qualifications; prepare owners manual data and installation drawings, and arrange for the availability of parts and materials. This development blitz was very successful, giving the flying qualities an important improvement, allowing a backseat stick and extending the allowable forward CG range. Three months later, four of the five Eze's at Oshkosh had the new rear wing ailerons. The other one was modified shortly afterward. I know of no Eze now flying with elevons.

Another problem that surfaced during the homebuilder's operational experience involved stall characteristics. The majority of airplanes we tested and reports from others indicated that the stall-proof flying qualities were indeed being realized. The airplanes could be flown in good stable flight indefinitely while at full aft stick. While holding full aft stick, the pilot could use power to climb or descend and could even sideslip with full controls without experiencing a stall break or departure from controlled flight. However, we had several reports from pilots who indicated they experienced divergent wing rocking at the stall or a roll-off when approaching minimum speed. These were occurring at aft CG but within the allowable CG range. Tests conducted by NASA with a VariEze model showed that the wing rock was caused by aft wing stall and that a small leading edge extension (cuff) on the back wing would eliminate it (Fig. 4-7, 4-8). In November, 1978, this departure (roll-off at minimum speed) occurred while demonstrating stalls in N4EZ. We then began a new flight test program to further investigate what was happening. We tufted the wing to observe airflow, and did high angle-of-attack tests with and without the wing cuffs. We found that the stall margin on the back wing without the cuffs was quite low at aft CG. Tolerances occurring during construction were enough to allow some airplanes to induce a stall on the back wing which results in a roll-off and loss of several hundred feet of altitude. Since tests showed this was eliminated

Fig. 4-6. Instrument panel on N4EZ.

Fig. 4-7. Leading edge cuffs show up on the wingtip of this VariEze built by Gerald Garner of Escondido, California. These cuffs are not movable.

when cuffs were installed, we then recommended them for all Eze's. Noting the large number of Eze's still flying without cuffs, it is seen that it is difficult to convince many owners to install them, particularly since most are very pleased with stall characteristics. However, because we found a departure stall on N4EZ only after nearly two years and hundreds of stalls, we still consider the cuffs to be a mandatory addition.

Operator experience indicated the need to increase the pitch stick forces, as many of the pilots were not comfortable with the sensitive controls. A modification introduced in July, 1978 increased the pitch control forces and greatly reduced the tendency for some pilots to overcontrol the Eze during their initial flights.

Performance variations experienced by the homebuilders have been large. Only a few airplanes can meet or beat the cruise speed listed in the Eze owners manual. The average Eze cruises 12 mph slower than the book. Small variations due to external additions, surface finish and fits can have a large effect on performance on an aircraft that has only 1.4 sq. ft. of equivalent flat plate drag area. While most Eze's can match the take-off and landing performance listed in the Manual, many suffer large losses due to improper propellers, wheel toe-in or pilot technique.

Many Eze's have experienced collapse of the nose gear when the downlock failed to retain the retraction mechanism. This problem was initially addressed with cautions and recommendations regarding rigging, checking for interferences and operating techniques. The problem, though never experienced with our prototypes, persisted on as many as 20 percent of the flying Eze's. The nose-gear collapse generally resulted in only minor damage to the skin under the nose, but, at times, major damage to the pilot's ego when he found his airplane on the nose closing a major runway. In late 1978, the nose gear retraction system was redesigned to a worm-gear mechanism which has eliminated the collapse problem and raised the gear-extension speed by 30 mph.

Several examples of failure or deterioration of landing gear attachments indicated that inadequate allowance was made in the original design to account for variances in workmanship and vibration loads due to wheel or brake imbalance. These resulted in recommended beef-ups of the attachments. These changes were not incorporated in the prototype.

There have been no airframe structural modifications required. The Eze's have a perfect record of airframe integrity, with no reported failures. Several incidents of violent flutter at high speeds have been experienced due to improper elevator mass balance. In each case the airframe was not damaged. (Flutter generally results in catastrophic airframe failure.) A canopy

Fig. 4-8. Wingtip cuffs are installed on Mike Dehate's VariEze. Mike is one of a group of San Diego, California builders who attend various fly-ins.

opening incident (Fig. 4-9) resulted in an Eze tumbling out of control while 25 mph over its maneuvering speed. No structural damage occurred. The pilot was quite complimentary about the airframe structural integrity. Enough experience has now been accumulated to assess the structural success of the composite structure. Approximately 15,000 flight hours have been accumulated at this writing. Environmental exposure had included the extremes of tropical salt water areas and Alaskan cold temperatures. Fred Keller's beautiful Eze which has flown to Oshkosh twice, is parked outside year-around at Anchorage. Several Eze's have now exceeded 500 flight hours. The oldest airframe is now nearly five years old. Considering the fact that the materials and methods are relatively new, that inspectors are in general inexperienced, and that there is an extremely wide variance in workmanship, the structural record is indeed impressive. We will, of course, continue to monitor the structural experience of the growing fleet of composite Eze's to assess their long-term integrity.

I am often asked what it is like to provide support for the over 3,000 builders and over 200 flyers. The answer is often challenging, often frustrating, often rewarding, but in general a lot of fun. My workload on support was reduced considerably in 1978 when Mike and Sally Melvill joined my brother Dick and myself here in RAF. Sally is our office manager. She handles the calls and letters, funneling them to Mike, Dick or myself. Mike has built a VariViggen and a VariEze and is in charge of builder support. Our quarterly newsletter provides the primary builder-support function and is the formal means to distribute any modifications and improvements. Since the VariEze program started in July, 1976, we have published a total of 14 newsletters which include a total of approximately 160,000 words!

Now for some statistics: In the last year there has been a new VariEze first flight approximately every three days. If the "average" builder calls us every five months for assistance and writes for help every five months, we get a phone call every 20 minutes of our working day (Monday through Saturday) and we get 24 letters per day. If only 1 percent of the builders have a problem with a given aspect of construction, thirty people will need help! As you can see, the builder support function is an enormous task. It is aided considerably by our requirement for all written correspondence to include a self-addressed stamped envelope and by our builder's cooperation in keeping calls brief.

Of course, there is no "average" builder. The exceptions are of interest. Two years ago, a fellow from Wyoming flew his Eze into Mojave to visit. He seemed unusually interested in looking over our prototype in great detail. We learned that he decided to build an Eze after seeing a photo in a magazine. He wrote to us only once—to order plans. He never called. He built and flight tested his Eze, then flew it to Mojave to visit RAF. Before his visit, he had never seen an Eze other than the one he built himself!

THE VARIEZE DEVELOPS

Rutan's initial announcement on the VariEze was prepared within a month of its first flight. The designer prepared the following rundown on the then-new concept and sent it along to the EAA's *Sport Aviation*. We have excerpted the following with permission.

The VariEze (Very-Easy) prototype has logged over 45 hours in its first three weeks of flying and performance is exceeding expectations. The aircraft was designed for maximum-cruise efficiency and an extremely long range. Major design features which make this possible are (1) light, yet rugged, structure using fiberglass in a composite form using rigid foam as core material, (2) compact tandem seating made possible by the broad CG range attained with the canard configuration, (3) modern airfoils—surface contour is maintained under load with the composite

Fig. 4-9. Canopy openings in flight have caused several VariEze accidents. A simple double lock system was developed by Rutan. Lee Roan of Temple City, California, shows the size of his canopy as he prepares to drop the nose of the VariEze for parking.

structure, (4) no control surfaces on the main wing, (5) Whitcomb-designed winglets reduce induced drag and halve the apparent parasite drag of the vertical fins (6) high aspect-ratio wings with distributed loading between wing and canard, (7) unique engine-cooling system that results in low drag, low oil temperature and longer accessory life.

The prototype has already demonstrated 70 miles per gallon with two people aboard—an unprecedented 140 miles per seat! That mileage is at the economy-cruise condition of 30 percent, 8,000', 135 mph TAS. At 75 percent power, it will deliver 48 mpg at approximately 185 TAS.

The low induced drag necessary for optimum cruise efficiency results in quite spectacular climb performance with the 62-hp Volkswagen engine: 2,000 fpm single place, and 1,200 fpm with two aboard and fuel for 1,000 miles. Empty weight is 390 lbs., normal gross is 890 lbs., which includes full fuel, two 6'4" people and two specially-designed suitcases.

The prototype was built in only 2 ½ months; thanks, of course, to Carolyn, who handled the VariViggen program, Phil Rathbun, who built the machined parts and Gary Morris, who donated most of his evenings to help with the glass lay-ups and finishing. That time included some tooling work also and we expect that a plans-built project can be completed in less than 350 man hours using supplied components. Components to be supplied include a canopy, landing gear (molded fiberglass), molded foam seat/bulkhead, cowling and wing spar/center section which will provide a "one-pin" wing removal similar to European sailplanes.

Flight testing is now being conducted to test modifications required to correct a canard flutter experienced with the original configuration. The present plan is for my brother, Dick Rutan, to fly the prototype to Oshkosh non-stop (1800 miles) with a 25-gallon fuel tank in the rear seat and, if time permits, attempt to set a new world, closed-course distance record during the '75 Oshkosh Convention.

This winter I plan to build another VariEze, using the manufactured parts intended for the homebuilt market, in order to fully test the supplied components and further simplify the basic structure. Plans and parts will not be sold until the entire development program is completed and parts are on the shelf. I expect this will be early in 1976.

True to Burt's fast-moving plans, the initial VariEze went non-stop to Oshkosh in 1975 and proceeded to make its name in the record books (Fig. 4-10). EAA *Sport Aviation* Editor Jack Cox thrilled the thousands of homebuilt enthusiasts with his report of the record flight. In part, here's what he had to say in an article that was written right at deadline following the VariEze's initial record flight. .

VARIEZE . . . FOR THE RECORD

5:30 is a brutal hour in the morning for anyone to be up and about after a week of endless tramping up and down the aircraft display lines at Oshkosh. Yet, a score or so of us have summoned the strength from somewhere and now find ourselves huddled around this pale apparition of an airplane, intently watching (Fig. 4-11).

Harold Best-Devereux, the NA/F.A.I. Official Observer for all that will hopefully transpire this day, jots down the numbers. Harold's crisp British accent snaps everyone back to the matter at hand.

"Now, gentlemen, will you please step back—completely away from the aircraft?"

As the rest of us back off a few grudging steps, Harold strides forward, reaches deep into the rear cockpit of the VariEze—behind the haulking fiberglass auxiliary fuel tank—and starts the barograph (Fig. 4-12). Then he proceeds to tape over the filler caps, affixing his initials to each with a flourish worthy of his station and

Fig. 4-10. Dick Rutan arrives at Oshkosh in 1975. Burt is there to greet him (courtesy Don Dwiggins).

Fig. 4-11. Silhouetted against the first light of dawn, Burt Rutan reaches for the propeller to set the day's events into motion (courtesy EAA by Jack Cox).

the occasion. Taking all this in with a slight bemused expression on his face is pilot Dick Rutan, waiting calmly in the front seat, already strapped in and ready to go. Dick is the only one in the crowd who really looks like he is up to the occasion. Decked out in a powder blue turtle neck pullover, he is downright dapper as compared to the rest of us who have the disheveled appearance of a bunch of sleepyheads who have just been rousted out of bed . . . which, of course, is exactly the case.

Draped over each of Dick's shoulders are several stout strings each attached to some unseen object behind his seat back—plastic bags containing a couple of Baby Ruth candy bars, a package of cheese and crackers, three pull-top cans containing Beenie-Weenies, chicken gumbo and chocolate pudding, a Chap Stick, Rolaids and some aspirin and a couple of plastic bottles full of water laced with just a dash of lemon juice, ¾ of a gallon in all. One string leads to a large, empty plastic bottle.

Not exactly an Apollo life support system, but simple, effective and, most important, lightweight.

A short exchange of pleasantries between Dick and Harold ends with a groping with the canopy support rod and a lowering and locking of the plexiglass bubble through which, if all goes well, Dick's only sensory contact with the rest of us will be possible for the next 12 to 14 hours, except for intermittent use of his battery-powered Escort 110 radio.

With brother Dick properly and officially encapsulated in the VariEze, Burt Rutan takes charge.

"Okay, let's have lots of hands under the canard and the main wing—anyplace except the control surface on the canard—and s-l-o-w-l-y ease forward off the scales and down the ramps.

"All together, now . . . lift!"

Effortlessly, the tiny bird, its pilot and 279 pounds of gasoline are palm-powered up, forward and gently down on the taxiway . . . its first "flight" of the day a total success.

Now Burt assumes the position at the rear of the craft and addresses the Monnett VW and Ted Hendrickson prop.

"Make it hot."

Flip . . . flip . . . flip. Come on you little Wolfsburg prima donna—this is no time for dramatic pauses! Flip . . . flip . . . broooom!

"Okay, just as a precaution, let's walk him out to the end of the runway. Keep the nosewheel light over the bumps and tar strips."

Burt and John Monnett climbed aboard the VariViggen. Now they have fired it up and are taxiing along behind the VariEze, preparing to accompany it on the first lap of this attempt to break Ed Lesher's World's Closed Course Distance Record of 1554.297 miles set back in 1970.

Dick's Air Force training makes him a believer in airspeed above all else . . . he levels out just after lift-off and accelerates in ground effect right down to the end of the runway. As Dick smoothly brings up the VariEze's nose, the rate of climb is far in excess of what even the new long-winged VariViggen is capable. I can't believe it; that little son-of-a-gun is carrying over 126 pounds more than its own empty weight, propelled by a 1700cc VW.

Leveling off, Dick throttles back to his programmed rpms and begins cruising up the shore of Lake Winnebago, heading for his

Fig. 4-12. Dick Rutan, Burt Rutan and Harold Best-Devereux check the barograph to assure that it would operate properly (courtesy EAA by Dick Stouffer).

turn point at Menominee, Michigan. Only then can the VariViggen catch up and slide in under the VariEze for a look-see to determine if all is well in the engine compartment. Finally, both have disappeared and those of us on the ground drift back toward the Control Center trailer to sit out the expected hour and twenty-five minute lapping of the Oshkosh/Menominee course. If that VW continues to purr away, we can expect to spot that unmistakable VariEze profile overhead at about 7:20 or so.

Burt Rutan had flown east in the VariViggen a few days before, leaving his brother Dick with the task of flying the VariEze to the EAA Fly-In at Oshkosh, hopefully in one big hop. The little bird had nearly 100 hours of flying time on it when Burt left and all manner of flight testing, fuel consumption tests, etc., had been accomplished. All that was left was for Major (Lt. Colonel by the time you are reading this) Richard Rutan, USAF, Field Maintenance Squadron Commander of the 355th Tactical Fighter Wing at Davis-Monthan AFB near Tucson, to get away from his duties, get up to Mojave and blast off on Wednesday. Flying non-stop to Oshkosh would be the grand entrance of all times for a new homebuilt design, Burt and Dick had figured.

In the wee hours of morning, Howard Gann and other local EAA types strapped Dick in and fired up the 1834cc Barker VW . . . only to have oil come gushing out of the cowling. The start-up had ruptured the oil cooler. A quick decision was made to remove the cooler, plumb the system "straight" and attempt the flight anyway. This wasted a precious hour or so but still left just enough time to make Oshkosh by sundown—if winds were favorable.

Taking off with nearly 50 gallons aboard, Dick climbed to 7500 feet and headed east, accompanied the first 100 miles by Howard in his T-18. The route to be flown was a gentle curve out across the Sierrras, southern Nevada, through the heart of Utah, clipping the corners of Wyoming, Nebraska, South Dakota, Minnesota and, finally, a dash across the mid section of Wisconsin to Oshkosh. The course was selected partly because it overflew major interstate highways and partly because it allowed some pressure system flying that promised tailwinds.

Settling back in the semi-supine and super-comfortable seat, Dick could see nothing ahead except clear sailing—visibility was unlimited, all the gauges were in the green and he was indeed picking up a tailwind.

A ground speed check somewhere over Utah revealed that the tailwind was really picking up. Later checks showed that a full 45 minutes had been picked up. Call a pre-selected FSS that Burt will call later in the day to let him know that ETA at Oshkosh will be about 20 minutes before official sunset. What luck! This called for another Life Saver from the package taped to the side of the cockpit.

About this time Lady Luck turned her beneficent smile elsewhere. Over Nebraska the tailwinds became headwinds and the oil temperature began to rise. Over southwestern Minnesota the oil temperature and the oil pressure started shooting up. Dick headed for the nearest airport and shortly was rolling out on the runway at Worthington, Minnesota, after 8 hours and 50 minutes of non-stop, non-refueled flying, some 1500 miles out of Mojave.

A check of the engine revealed nothing that could be seen, except that most of the oil had been consumed. This would explain the rise in oil temperature, but what caused the oil pressure rise? After replenishing the oil supply, a run-up showed everything in the green again—and left a nagging suspicion that in the rush to remove the oil cooler that morning, maybe the oil had not been topped off before takeoff. Better call Burt and give him the bad news.

Burt, meanwhile, had been following the progress of the flight by calling in to the FSS stations along the route that he and Dick had agreed to use as "message drops." Dick was hardly on the ground before Burt knew about it and shortly the two were talking it all over by phone. After hearing about the puzzling oil pressure reading, Burt agreed Dick had made the wise decision, sparing himself and the VariEze to fly again another day.

After a night's rest, Dick flew on to Oshkosh the next morning to the most spectacular reception an aircraft has ever received at an EAA fly-in. Several PA announcements had been made the previous day keeping everyone informed of the VariEze's progress as it winged its way across the continent, and Burt had talked briefly at the evening program detailing the problem with the oil pressure. An announcement was made Thursday morning when the tower reported the VariEze entering the pattern and it appeared that everyone there was standing on the show line to witness the landing on Runway 18 at 8:40 a.m. There was no way to taxi in through such a multitude; the tiny aircraft had to be walked to its already roped off parking spot beside the VariViggen. There it was to be totally surrounded by huge crowds every minute of the daylight hours that it was on the ground.

A thorough check was immediately made of the engine and nothing could be found awry. After removing the long-range fuel tank from the rear cockpit, a test flight was made with this writer serving as an inadequate replacement for the weight of 35 gallons of fuel. Again, no problems.

On Friday, Dick took the bird out for some more flying and on landing, had to go around to avoid a slow plane rolling out long on the runway. Cranking the nose gear up and then right back down again for the second landing attempt, he apparently did not get the circulating ball system wound up to the stops, although it had felt to him it was "down and locked." On touchdown, the little panel-mounted crank started spinning wildly, slowly letting the nose right down on the pavement. A layer or two of fiberglass was ground off the nose gear leg and the bottom skin, but that was the extent of the damage. Repairs were made by Gary Morris right at the aircraft's parking spot—with a pair of scissors, a paint brush, a can of epoxy resin and strips of glass cloth. By evening, 7EZ was pronounced ready for the record attempt the following morning, Saturday, August 2.

All that day frantic preparations were being made elsewhere on the field for the record attempt. Several weeks earlier, Burt had made application to NAA, the U.S. F.A.I. affiliate, for the attempt. David Scott had been designated as the official NAA observer . . . and he had much to observe even before the first prop was turned. A barograph had to be smoked and sealed, scales had to be certified, turn point observers on the other end of the closed course had to be lined up, communication with the Oshkosh tower had to be coordinated, etc. Fortunately, Bill Turner also became available and pitched in with the legwork. Harold Best-Devereux, who was an old hand at this sort of thing, was there whenever he was needed.

Adding to the last-minute adrenalin level was the fact that the closed course was changed at the eleventh hour. All week the weather had been unseasonably hot for Wisconsin. Gulf moisture was being pumped up the back side of a titanic high pressure area stalled in the east, resulting in a really bad haze condition all over the midwest. The original closed course was to have been from the Oshkosh Omni to the Burlington, Wisconsin Omni—but the rather featureless Wisconsin landscape would make Burlington awfully hard to find groping through the atmospheric goop. The use of omni could not be depended upon because the VariEze's radio was powered only by a primary system consisting of an 8-amp gel cell and a secondary system consisting of a 2-amp motorcycle battery. Only intermittent use would be possible because the electronic instruments were also drawing off the power supply. The Barker engine was devoid of all but mags and a carb to keep it at a spare 138 pounds. Starters and generators simply meant less fuel, reasoned Burt. Being the Original Interstate/Railroad/Coastline Chicken Flyer, I suggested a course I have often flown: up the west shoreline of Lake Winnebago, over the freeway to the city of Green Bay and up the west shore of Green Bay to Menominee, Michigan and return—a 182-mile, no-sweat navigation run, even in marginal visibility. This met with everyone's approval.

Now, if the weather will cooperate . . .

Saturday morning started at 4:30 a.m. for those of us involved in the launch. A quick breakfast for some of us, none for others, and it's off to the airport, with one eye on the somewhat low overcast, the first since the fly-in started. The weathermen say some scattered showers and maybe a thunderbumper to fly around before the day is done, but ceilings should be VFR. The weighing, sealing of the tanks and barograph, etc., proceed under the direction of David Scott and presently the buzz of the VW is causing heads to peep-out through tent flaps in the campground.

As Dick taxied out, Burt ran by and yelled at me, "Jump in the back of the VariViggen and we'll pace him the first lap. Be back on the ground here at 7:30."

I dashed for my camera and hopped in behind Burt, who already had the Lycoming turning. Taxiing out to the end of 18, we lined up behind and to the left of the VariEze and followed him down the runway at a distance that left us some place to go in case he had to abort. It was difficult to see much of the VariEze's takeoff from the rear 'pit, but it was long and climb wasn't initiated until a real head of steam had been built up. We were already off and climbing, but when Dick started a climbing turn back to the northeast, we seemed to be settling as he zoomed up. Burt kept yelling something about how how he couldn't catch up, that we were at full throttle. Also, he was laughing a lot.

When Dick throttled back to 3075 rpm, we were gradually able to catch up and finally pulled alongside, indicating just over 130 mph. We then slid up under the VariEze for a look at the engine and were greeted by a chilling sight . . . a brown streak, at least two inches wide, streaming back from the air inlet all the way back to the prop hub. Oil!

Some animated radio conversation between Burt and Dick resulted in the hopeful conclusion that perhaps the oil was merely residual spillage in the cowling, because all the instrument readings were comfortably in the green. But we would keep a close watch the remainder of the lap to determine if the streak became wider or darker.

On we speed, over Neenah, Menasha, past Appleton, over Kaukauna, up U. S. 41 to Green Bay. The overcast begins to break up and by the time we are approaching Menominee, shafts of sunlight are creating luminous, shimmering pools on the otherwise drab green surface of the bay.

Sweeping around the easily spotted airport, we see Paul Schultz, Joe Gypp and others spreading a white sheet near the Enstrom helipad, indicating a confirmation of our pass. Burt has also received the good word via Unicom.

Turn completed, we head southwest toward Oshkosh. Sliding in under the VariEze for perhaps the tenth time, we can detect no change in the size or color of that ugly streak on its belly. As the landscape slides so rapidly beneath us, it is easy to believe our assumption that the oil is spillage—merely spillage.

Over the city of Green Bay, the overcast becomes solid again and as we proceed down towards the north shore of the Lake Winnebago the area ahead of us darkens dramatically. By the time we are over Kaukauna and Little Chute, we are in light rain and the ceiling and ground fog that has materialized from nowhere are ominously close to what will be an illegal merger for Burt and me in the VariViggen. We push on for a couple of minutes, but when the near all-white VariEze starts pulling momentary disappearing acts on us, we know we've been had. A quick call to tell Dick to follow the dual lane road south rather than following a coastline he sees to the west—that's Lake Butte Des Morts—then Burt takes full advantage of the Viggen's turn-on-a-pin-head capability and we are headed back to Green Bay airport. Somehow Dick gropes his way through to Wittman Field, gets confirmation on his turn and starts back north behind us. One lap completed.

Burt and I race the rapidly advancing line of crud back to Green Bay's Austin Straubel Field, land and dash into the FSS to see what the heck has gotten the weather god's bowels in such an uproar. About 15 minutes later, I stepped outside into a light sprinkle and was greeted by, "Hey, where were you guys when I needed you?"

To my utter astonishment, I turned to see Dick Rutan striding up the walk.

"Weather?"

"No, just blew the engine about 20 miles north of here. Made it back by pumping the extra oil we installed last night. Dead-sticked in here."

"Dead stick . . . with all that fuel on board?"

"Had to, the oil pressure was reading zilch. Thought you guys would come running out to help . . . had to push the little beast in to the ramp from out there in the middle of the runway."

We walked back into the FSS, turned a corner and confronted Burt. "Thunderstruck" is a pretty good adjective to describe the look on his face when he saw Dick. Out again into what had now become a light shower of rain, we trudged out of the VariEze and hunkered down to view the oil soaked belly.

All the effort, all those people at Oshkosh and Menominee who have helped out—and here we stand watching oil drip-dripping onto the pavement.

"Well, the weather probably would have zapped the flight anyway.

With the downpour getting worse by the minute, some kind gentleman drove out and invited us to push the VariEze into his hangar, which we gratefully accepted. Pulling off the cowling we find . . . absolutely nothing. No gaping hole in the case where a rod has smashed its way out, no ruptured hoses, nothing. More probing leads to the conclusion that the VW had spilled its oil out the number 3 cylinder, but it was impossible to say from what specific point because the entire lower side was covered with the stuff.

I suppose for some this would have been sack cloth and ashes time, but not so with Burt and Dick. Conversation immediately turned to where and how they could get a new engine and be ready to go *Monday morning*! With all the engines at Oshkosh, there's bound to be one that can be used, is the reasoning. It can be installed tonight, test flown tomorrow and be ready to go Monday morning.

By various means, all of us got back to Oshkosh during early afternoon.

Golda had John Monnett waiting for me when I walked in the door at Press Headquarters. Sure, he had a brand new engine in his booth, ready to bolt on . . . but it's brand new, no run-in time, the mags would have to be timed and it had a Posa injector carb. He had a better idea. An hour or so later we caught up to Burt who had finally managed to get the VariViggen back to Wittman Field. Right there in the middle of the busy display building floor, John laid a deal on him that was impossible to refuse.

"I'll send a couple of my friends down to Chicago tonight, have them remove the engine from my pranged Sonerai, fly it back, and my crew will work all night installing it in the VariEze so you can start test flying tomorrow. With a new engine, any new engine, you can't be sure what you have for the first 20-25 hours. With my engine, I know what you've got."

In late afternoon the VariEze arrived on a loaned trailer. After no little searching through the vast EAA grounds, Sonerai builders Charlie Terry of Long Island and Vance Graebner were located and immediately dispatched to DuPage County Airport to remove the engine from John's bent bird. It was after midnight when they returned with the vital organ. John and Mike Core would spend the remainder of the night transplanting it in the pallid body of the VariEze.

By sunup two very weary bug doctors had completed their work and were ready to look for some breakfast and a couple of beds. Burt could handle the final closure, cleanup and bandaging. Considering the drama of the past 24 hours, it was almost disappointing when the engine simply fired up and ran like it was supposed to; well, almost. It ran, but Burt was not happy with the characteristics of the Posa injector, so off with the cowling, off with the Posa and on with the Barker engine's float carburetor. Whoops! the intake plumbing doesn't fit, and it's Sunday. Probably the only place in the U. S. that day with all sorts of aircraft hoses and hardware for sale was the EAA Fly-In. Within 30 hours of Dick's dead-stick landing at Green Bay, the VariEze was winging its way around the fly-by pattern at Oshkosh.

One last dollop of adrenalin remained to be squeezed out of the situation. After landing, Burt eased off the runway, came to a stop, shut down and climbed out to inspect the nose gear leg. The earlier fix had not been enough—a crack had developed. No big deal, however, as the repair, including an additional wrap with glass cloth, took only an hour or so. Most of that was curing time.

At dusk all that could be done had been done, so everyone involved headed for bed. A 4:30 wake-up call would be much harder to take this time around. Come morning the same cast of characters would greet the rising sun.

And that's how we came to where we are . . . standing around or absently walking over to look at a couple of homebuilts, waiting, watching for the VariEze to return. then, finally, there it was—the

VariViggen. Can't see the VariEze yet, but the "mothership" must be leading it in. Yep, there it is! What a beautiful sight!

After swinging wide around the Oshkosh tower where Harold is standing by to confirm the turn, Burt peels off and enters the landing pattern. Good sign! The VariEze must be okay if Burt is letting him head back for Menominee.

When Burt and John taxi in, we descend upon them for work on the VariEze, and they report that all seems well. Now they join the ranks of the watchers and waiters. We busy ourselves with the statistics of the first lap: Airborne at 5:55 a.m., Over Oshkosh Tower at 7:20 according to Harold Best-Devereux's watch—an hour and 25 minutes to cover 182 miles.

That's 128.5 mph and includes the climb-out from Wittman Field. Before landing, Burt has gotten fuel consumption numbers from Dick by radio and he seems concerned, but he isn't saying much.

"We'll see how it looks on the end of the next lap."

Lap two ended with Harold Best-Devereux's, "Mark 8:44." That was one hour and 24 minutes—130 mph.

"Too fast," says Burt.

He uses the VariViggen's radio to order a power reduction-from 3075 to 3050. His brow knits a little deeper when he hears the fuel consumption figure for lap two.

"Ladies and gentlemen, the Oshkosh Tower has just established radio contact with the VariEze. If you will look to the northest, you will soon see this aircraft completing its third lap."

"Ladies and gentlemen, the VariEze is again approaching Oshkosh. Pilot Dick Rutan is completing lap 4 and will be beginning lap 5. When he passes over the Oshkosh Tower, the VariEze will have passed the halfway point toward breaking Ed Lesher's record. Nine laps are required to set a new mark."

"Mark 11:40" a one-hour, 27-minute lap—125.5 mph. Burt doesn't look quite as worried over the fuel situation.

"Mark 1:09" One hour, twenty-nine minutes—122.5 mph. Funny, the mid laps seem to be going past faster than at the beginning. Complacency? . . . or is hunger dulling the senses? It's been eight hours now since breakfast.

""The VariEze is inbound again. This will be the completion of lap 6. At the turn, the VariEze will have covered 1092 miles. This is the first time a world's record has been attempted at an EAA Fly-In. We invite everyone to stick around this evening to greet Dick Rutan when he completes the flight."

Harold's mark had caught Dick rounding the Oshkosh tower at 2:36 p.m., an hour and twenty-seven minute time for lap 6. Same as lap 4. Obviously, wind is not a factor today.

"Mark 4:07. Two more laps for the record, gentlemen." Hmmm, that's an hour thirty-one—120 mph. A check with Burt reveals that, yes, he did slow Dick down again to 2950 rpms. Fuel consumption?

"Yeah, it looks like we are burning a little more than we expected. Don't think we will be able to go the extra laps we planned. Running too slow now . . . but the record looks okay. Know what? I don't think we had the tanks completely full at takeoff. I couldn't believe the consumption on the first lap, but has settled down some now."

So that's what was on his mind.

"Mark 5:37."

"Ladies and gentlemen, the VariEze has now completed lap 8 and has started the record lap. If all goes well, the aircraft will return over Wittman Field at just after 7:00 p.m. at that point, Dick Rutan will have flown 1638 miles, 83.7 miles further than Ed Lesher's 1970 record."

Decision time! While Arv Olson is keeping the crowd informed over the PA, Burt is busy taking data from Dick via the VariViggen's radio—speed, fuel remaining, temperatures, pressures—the decision has to be made now on trying lap 10 because now the race is also with the sun. The VariEze is not equipped with lights and a tenth lap at the present lap speeds would get Dick back around 8:30, after official sunset. It there enough fuel left to speed up?

Lap 8 took an hour and a half even . . . it sure seemed longer than that: 121 mph.

Look at the people who are beginning to gather around the Comm Center.

"He's coming in this time," somebody yells. A dash to the Comm Center confirms it. Dick has decided call it quits at the end of lap 9; the fuel remaining is such that 10 laps would be slicing things too thin. There's a technicality that has to be kept in mind in these closed course record attempts—you have to land back at the same airport from which you started, otherwise all goes down the tubes. Dick has figured his fuel at the turn at Menominee and has told them via Unicom to call us regarding his decision.

"Ladies and gentlemen, the VariEze will land at the conclusion of the 9th lap, setting a new world's record. When the aircraft

Fig. 4-13. We did it! Paul and Audrey Poberezny, Dick and Burt Rutan, Harold Best-Devereux and Bill Turner after the record flight (courtesy EAA by Dick Stouffer).

lands, everyone is asked to stay back behind the showline barriers. For the record to be official, Harold Best-Devereux, the official observer, must check the fuel tank seals and remove the barograph before the plane is disturbed.

Harold's "mark" comes at 6:58 p.m., officially ending the course time. That is a 1:21 lap, the fastest of the day. Just over 134 mph. We will let the tower mark his official touchdown time and figure his total time in the air from that.

Just over 13 hours aloft is not enough to cool off Dick's enthusiasm before landing on 18! He says the VariEze cockpit is the most comfortable he's ever sat in; must be true.

This time around he has the nose gear cranked down; hope it's locked. That's it, he's down! He's done it!

Dick taxies back up the side of the runway and turns down the EAA access. Harold is out of the car now and is giving his the "cut" sign. Dick brakes to a stop and is lifting the canopy. He is greeted by a resounding cheer from his fellow EAAers (Fig. 4-13). Harold gives his a fast handshake and proceeds to dive into the rear cockpit for the barograph. Presently he emerges and hoists it over his head like a trophy won. More applause.

Dick pulls out his can of chocolate pudding to show what provisions he has left and immediately it is requested by an admirer, who also wants it autographed after the prize is his. This starts a frantic round of autograph signing by both Dick and Burt.

Dick was in the air a total of 13 hours 8 minutes and 45 seconds. The tower officially had him down at 7:03.45 C.D.T. The nine-lap course distance was 1638 miles. Burt Rutan finally figured that the VariEze had taken off with 46.5 gallons of fuel on board; 6.3

gallons remained when the flight was completed; so 40.2 gallons were consumed in the 13-plus hours. This figures to just over 3.1 gallons per hour for the day's flying. One pint of oil was used by the Monnett VW. The average speed had been 125.5 mph. These were tremendous figures for any small airplane, but more impressive when one reflects that the construction of the aircraft was started the last of January of this year and that it did not fly for the first time until May 21. The months and years ahead will see the effects of the shock waves that are even now rippling out through the aviation world. We suspect they will be profound. Certainly it can be said, no homebuilt design—or factory design—has made such a spectacular start as the VariEze. It took a lot of help from Burt's friends to get the first record, and he is grateful; but to Burt must go the credit for daring to be different in the design of this aircraft by asking so much of it so soon . . . in full view of so many people.

BUILDERS' EXPERIENCES—FACTS, FRUSTRATIONS AND FOUL-UPS!

The trials and tribulations of VariEze builders differ somewhat from those of VariViggen, Quickie and Long-EZ fabricators in that the VariEze was the first composite construction homebuilt to achieve true popularity. Thus, many of the problems were those of first time pioneers.

A Mature Builder Speaks His Mind

We met VariEze builder Norm Spitzer (Fig. 4-14) on a hot summer morning in the RAF office in Mojave. We'd met him before, a long, long time ago on a rice paddy in the middle of the

Fig. 4-14. Norm Spitzer, WW II pilot, with his partially-completed VariEze, shown on the porch of his second-story apartment. "It has rekindled a new interest and joy in aviation which I thought I had lost," said Spitzer (courtesy Jean Spitzer).

Upper Assam jungle in those nearly-forgotten days when both of us were flying C-46's over the Burma Hump. But enough of the war stories.

Norm's VariEze, which he calls "The Rockpile Express" in memory of the India-China-Burma "rockpile" of the World War II days, was completed and flying when we ran into him at Mojave. He was close to having flown his time out in the Central California Bay area near Berkeley and was savoring the idea of taking his new plane on its first long cross-country trip. We asked Norm to share some of his experiences in developing his project. At first, he didn't want his name used because he said that he wasn't reaching for notoriety, but we pointed out that the material was more forceful with a real, live, breathing builder in the background, so here 'tis.

"You asked what prompted me to start the VariEze. After forty years of flying, I found that I was becoming bored with the stock complacent Wichita type of aircraft. I had become more and more interested in the home-building movement. The first information on the VariEze captured my imagination with its efficiency, type of construction, and most of all its magnificent appearance. And then, a very close friend (we had been Cadets together and subsequently, fishing and hunting partners) died unexpectedly. He had taken an early retirement from the airlines to do all the things he had put off and shortly after that he was gone. My wife pointed out that I had been talking of building an airplane and not to wait too long, so the decision—thanks to her—was to start the VariEze.

"We drove to Aircraft Spruce and brought back the kit in our station wagon. Plans were acquired and my neighbor, Gene Cartwright, recently retired from the newspapers, volunteered to assist.

"Several other builders were located locally and wonderful telephone conversations and visits were initiated with discussions of construction methods, sources of materials, potential performances, etc.

"The one frustrating aspect was the constant changes that came during the early period of construction. I had the control system finished when ailerons were added; the fuselage when the dive brake was added; the gear when the retraction method was revised, etc., etc. It got to the point where the quarterly newsletter was received with dread as invariably the changes would set back the completion date by months.

"I had built the ship in a bedroom on the second floor, and as each unit was finished, it had to be stored elsewhere because of space requirements. Soon wings were stacked in the entrance way, boxes of parts in the living room—with urethane dust everywhere. Meanwhile, the mailman and UPS brought a constant stream of strange boxes to add to the modifications, such as rejecting and rebuilding items, and the one main tragedy of breaking the canopy after its completion.

"Then comes the subject of visitors. The knowledgeable ones are always welcome, with many delightful hours spent displaying plans, parts, and learning from them; then, there are the others with quick advice without knowledge. Actually, the visitor situation is very good preparation and training for your first fly-in when your plane is inspected by a horde of hard-eyed, doubting fellow VariEze builders. On one occasion after looking at the alternator, instruments, radio, omni transponder, etc., one such type said, with his voice dripping with disbelief, 'How did you get it so light?' Thanks to the visitor training period, I was able to instantly answer, 'Oh, I left the main spar out!'

"Finally, I took the ship out to the airport ready to fly. Four months later, it did fly, but not before another sad occurrence. My good friend and helper, Gene Cartwright, died of a heart attack. I had decided to fly it when both the ship and I were ready and so did not invite any observers. However, the two local VariEze pilots saw me taxi out and were on hand for the landing. Now if you read *Sport Aviation*, you know the first flight is supposed to be, at the very least, some type of religious experience. In my case, this was far from true. The ship was wing heavy and I was mad at myself to have erred, and I guess after three years of construction I was let down. My friends insisted on congratulations and retiring for a ceremonial drink. I did appreciate their thoughtfulness—I never did find the source of the trim problem and finally corrected it with a fixed tab.

"Only as the flight test program progressed did I really begin to appreciate the wonderful experience of developing your own airplane. I have plans for many changes and experiments down the road, and this ship will fullfill my aeronautical appetite and curiosity for many years to come. It has rekindled a new interest and joy in aviation which I thought I had lost. My wife Jean, who has contributed so much in my life, has taken a complete part in the entire endeavor, and I feel sorry for those who start such a project without that support. Finally, the building of an aircraft introduces you to some great people. Buchanan Field's other two VariEze builders, Lyle Powell and Carlos Amspocker, fall in that category.

"And finally for me, building this ship was never VariEze. But Carlos, a retired chemical engineer and ex-Marine pilot, puts it very well when after flying, he sits in a chair, lights up a cigar, looks at his VariEze parked in front on the hangar, and says, 'I can't believe it's all mine and that I actually built it!' "

Trio With 500 Hours Each

The first three VariEze builders to top the 500-hour mark all live in California. They are Ed Hamlin of Rocklin; Les Faus of Van Nuys, and Dr. Don Shupe of LaVerne (Fig. 4-15). The trio met and discussed their experiences with a tape recorder running, using a series of questions as a format. Some of their comments and stories, gleaned from an hour and a half of tape, give an interesting insight to these builders of VariEzes.

How did you first become interested in the VariEze?

All three credit an article published in *Air Progress* magazine and all were impressed with the mileage and speed of the two-placer. Les Faus started out looking at the original BD-5 and planned to purchase a production model in partnership.

Les: That got me into flying, because the fellow I went into partnership with said I wasn't going to fly his half until I got some time in. I hadn't flown 'til then. Then the VariEze came up about that time, and it looked like it was going to be a long time until the BD-5 was done, so I got interested in the VariEze and went ahead with it.

Ed: A friend came by and wanted some help in building a seaplane in my garage because I had the better shop. My friend was going to buy the materials and I'd go ahead and build it. This was back in the summer of '75. I became disillusioned because there were only five places to land seaplanes in California. Then I saw the *Air Progress* article and said to myself, "That's the airplane I've got to build." I was familiar with the fiberglass material. At that time, I owned a 1/6 interest in a Grumman Trainer but I couldn't afford to own a certificated plane because of maintenance and other problems.

Don: After I saw the article, I went around talking with friends, discussing the specifications that Burt had put on building time and simple construction style. The price tag, including engine, seemed to be within my financial budget, and I had had some experience in fiberglass on boats. It seemed to be the way to go. I had owned a Cessna 150 on a leaseback and it ate me alive. It cost about $4,000 a year to own it and rent it out. I couldn't afford that kind of plane either.

Les: I was influenced by the fact that you could get a higher performance airplane for a lot cheaper than a factory one. I did get into a partnership with a Cessna 150 for a while just to build time.

How long did it take you to build?

Don: I started in the summer when the plans came out—one of the first sets of plans. I worked steadily for two full summers— about 80 hours a week, and then another full year; it first flew in December '77.

Les: I was one of the first ones to get plans. It took me 10 months to finish and the cost turned out to be about $7,500.

Ed: I got started in the summer of '76 and my airplane first flew in March, 1978—18 months. I spent a fair part of the summer of '77 and then laid off of it because I was burned out. Sacramento Valley was hot and it was hard to work. I couldn't keep the shop cool.

What kind of building problems did you have?

Don: Every kind possible. Specifically, I built the canard first to specifications and did all the measurements as closely as possible. I got done and found that the elevators didn't have sufficient travel because the hole they were supposed to go in wasn't big enough. That was because the pivot arms were down too far, so I had to cut them all out and put them on again. That was the first thing that I did twice. I did dozens of things twice after that because I was one of the first ones to check out the plans. There were lots of errors in the plans. During the time I built, I found over 30 major errors in the plans for the first time that weren't reported

Fig. 4-15. First VariEze builders to log 500 hours each in their canard homebuilts are Dr. Don Shupe, left, Les Faus and Ed Hamlin.

123

anywhere else. Some of them I found before I made mistakes; but a lot of them I found after I already had the part completed, so I had to do a lot of things twice. That caused lots of frustration and slowed the project down tremendously.

Ed: Well, you had no practical building experience with this type of construction, did you? Actually, there were very few things I had to do twice. I rebuilt one of the back bulkheads simply because I mis-measured one and had it half completed when I finally found my mistake. Other than that, things went fairly smoothly, but I had an extensive background in building fiberglass and foam radio-controlled model airplanes which helped me a great deal in the construction of the airplane—plus, my building partner, Dick Kennedy, had experience working on large airplanes. We had several other homebuilders who were very helpful in our EAA Chapter, and when a problem arose, we could go to several people to try to figure out how to solve it. These people were relatively close at hand. So things really went relatively smoothly.

One of the things that really slowed our project down was that we made all our own metal parts, including wing fittings, which included the unique experience of buying bars of 4130 steel and finding machinists that would turn out our taper pins, finding a tapered reamer of the proper angle and specifications so we could ream the holes. But all totaled, we probably made about eight or nine sets of wing fittings. That is the thing that really slowed my project down.

Don: What slowed me down was that being one of the first people in the area to build, Brock didn't have a lot of the parts done that I couldn't make. I had to wait a long time for wing fittings and had to go ahead with the fuselage before I could do the wing fittings—which turned out to be just fine.

But one of the things that was a major problem was stuff like—Burt said you were supposed to mix the slurry that was supposed to go over the foam to a fairly thick consistency. The ratio he suggested was 50 percent microspheres, or something like that. The mixture I would get when I would mix that would be a gooey mixture that simply would not spread over the foam, and so I would get these chunks of slurry-type stuff that wouldn't even out and I would be starting with a nice smooth piece of foam and by the time I got done putting the damn slurry over it, it would be all chunky and lumpy. And by that time—he told us to use fast epoxy—and I had a 75° room, and the stuff would start setting up. If you didn't get your glass on there quickly, you'd have a horrible mess. Then, later on, I discovered that people were using a much thinner slurry that spread out very smoothly and made all the difference in the world.

But see, I didn't have any help. I didn't have anybody to go to to ask how to do things except to call Burt. The first four or five months I was working on it, I had $50 per month phone bills just calling Mojave. They were very helpful, but lots of times I didn't even know what kind of questions to ask. I knew I was having trouble, but I didn't know what to ask.

My layups were taking three times longer than what he said in the book. But by the time I got done with the last wing layup—he said it should take two hours—six of us got it done in two hours. But it took three other wing layups to get to that point.

Another major problem I had was that I didn't know anyone who knew how to cut foam. So when we cut the wings and the canards, the guy who helped me and I were cutting the first piece of foam we'd ever cut in our lives. The wire was too hot—we overburned, and the wings required major surgery to save them. But we tried to salvage them and as a result the wings are just super rough.

Ed: I had the experience of cutting hundreds of sets of foam cores, because once I got involved with a guy who was running a hobby shop. I cut him hundreds of sets of foam for links and delivered them to him as a part of a project. We also had the foam cutting equipment because a friend of mine from the model airplane days had built a very nice rheostat-controlled foam cutter.

Les: The only building problem I had was epoxy poisoning. The airplane went together without any problem, but the epoxy poisoning would keep me from working two to three months at a time. Even now, if I go out and wash car parts with paint thinner, I'll break out the same as I would with epoxy. With the epoxy, it's definitely fumes; with the thinners, I don't know. I don't need to touch the epoxy at all—all I have to do is pour the hardener from one can to another and I break out. I used a respirator after it hit me. I didn't take the precautions I should have like I was warned. I also was working in a closed garage trying to keep the heat up for the epoxy, so I was in a heavy concentration of fumes to start out. But I had most of the airframe done in six weeks. It was a struggle from there to finish the rest of the stuff.

Ed: Six weeks sounds really fast.

Les: Well, it was a week each for the wings, about a week for the canards, and the rest of the time for the fuselage boxes. I was pretty well done with the nose, but I needed fuel tanks on it and the canopy.

Ed: How long were you stopped that first time?

Les: I didn't work on it for a month to get over the epoxy poisoning. Then I started dabbling a little at a time to see what I could tolerate and what I couldn't, and how I could work with it.

How much time did the project take?

Ed: I would go out night after night after work, around 6:00, and work until 2:00 or 3:00 a.m. Weekends I worked straight through with only four or five hours sleep. I kept at it constantly for 1 ½ years; 2,000-2,500 hours; it could go as high as 3,000 hours.

Don: That's about equal with my figures, which I believe to be fairly accurate. They indicated 2500 hours minimum. The problem was that I would go and sit down with the plans and stare at that damn thing for hours before I would get up and actually do something. I'm also including that time that I was in the room with the airplane where I was trying to figure out what I would do next. That's why I believe Burt's figures are so much lower in the minimum hours—he didn't have to spend that particular time. For me it had to be probably one quarter of all the time—just trying to figure how to start the next part.

Les: Of course, being one of the first, I had to make all changes that Burt came up with, including the spoilers.

Ed: You built the spoilers?

Les: Yeah.

Ed: And threw them away?

Les: Yes, it flew the first time with the spoiler system. Of course, when I took it up to Mojave, Burt said throw that garbage away, it's no good. He had just come out with the plans for the ailerons about the time that I showed up.

Ed: Did the fact that you had to go back and do all this retrofitting bother you?

Les: Only from the epoxy standpoint. After it flew the first time and we decided it needed ailerons, I put it in the garage and let it sit for three months before I got enough courage to go back and get involved with the epoxy again.

My background as a machinist and in building experimental kinds of things lends itself a lot better than most guys that are building. I've seen how much trouble some of the fellows have had when they don't know how to read blueprints; they haven't had any experience in that or in building things for themseves. I makes a vast difference in your approach to things.

Did you have any problems with FAA inspection?

Les: No problems. The inspector I had said, "It's your fanny—not mine."

Ed: I felt the FAA, at least my GADO, was possibly very apprehensive in the very beginning of the whole concept of the airplane—it's so unusual.

Don: The inspectors that looked at my airplane were seeing the first Eze they had ever seen, and so I had to run through and explain to them how it was built and what was involved. One of the guys went back, and he was very upset because he couldn't see the spar inside the wing. What they decided at one point in time is that they were going to have people call me to look at the spar box for the wing before the wing was assembled. As it turned out, the way Burt described building that thing originally was that you were supposed to stick the wing together and put the foam back on the spar while the spar was wet. That meant they had to be there at that instant if they were going to look at it right from the time we layed it up while it was still wet. We argued round and round about that because they wanted to look at it. They finally conceded that it was not a practical thing to do. Other than that, there were no problems with the FAA.

What was your level of piloting experience prior to flying your VariEze?

Les had about 200 hours in Cessnas and just before flying his VariEze, he took an hour in a Citabria. Don had 350-400 hours, almost entirely in Cessna 150's, but including 15 hours high-performance retractable time. Ed had about 300 hours, with a majority of time in high-performance aircraft—Beech Sierra and Beech Bonanza. He also had a good portion of time in a 160 Grumman trainer that had similarities to the Eze in high wing loading.

What kind of preparation did you make for your first flight?

Les: Only special preparation was flying the Citabria for an hour and getting familiar with that. Here again, I'm used to running machinery and I have a light touch on things, which makes a bit of a difference compared to somebody who doesn't use their hands.

Burt made the first flight in my Eze and then I made the next flight after the ailerons were put on. I knew the airframe was straight, so I was going to do a lift off and the runway turned out to be too short by the time I got it up fast enough to get it lifted off, so I just poked it and went. I learned to fly it in the air and came back and made a couple of passes for landing. The major thing I found wrong

Fig. 4-16. Desert weather at the Mojave Airport and environs can become interesting. This is the area where Les Faus and many other builders "fly off" their first 50 hours. Summer thunderheads are shown building over the area.

was Burt had the speeds too low for takeoff and landing. That's right at my stall speed: mine stalls about ten miles an hour faster than his figures.

So I was having trouble controlling it at the low speeds, because the wing rocks quite badly at stall. My approaches were too slow and my nose too high. After about four landings, an observer told me it would look better if I went a little faster. I upped the speeds 10 mph and everything went beautifully from there on. In three or four flights I was really comfortable in the airplane. It's more like a sports car; it goes where you point it.

Don: Really, I didn't make any preparation for test flying. I was working on the plane so much and I had very little money. It was just not practical to get out and fly a taildragger or anything else. I did extensive taxiing that I'm sure was not good for the airplane engine, but it was really good for me. I was down at Chino on a long smooth runway and did many, many runs before I finally got it off the ground. I knew the airplane was rigged properly and that it was balanced. As it turned out I required minimal trim. When it went off, it was all ready to go.

Ed: I didn't actually do any preparation. I was flying fairly actively at the time.

What problems did you have during test flights?

Les: Everything went beautifully during the test flights. The only problem I had was pilot error—I didn't put the gear down once. I flew 50 hours for certification—right on the money when I got it signed off—in a little over a month. I did all my flying around Mojave (Fig. 4-16); some days I got in eight hours of flying. You have 25 miles for a test area (Fig. 4-17). There were three of us flying the test area. The other two were a BD-4 and a Grumman TR2. The Grumman was a composite aircraft—two wrecked ones that had been glued back together. He had to fly off 15 hours and it took him about four months to get the 15 hours.

Don: I had been practicing so it was lifting off a couple of inches, and I could feel that it was stable. On this one particular day, I used full power toward the middle of the runway and instead of holding the stick down, I let the stick neutralize itself. If centered itself and the plane just came right off and it went up about 20-25 feet. It scared the hell out of me and I didn't know what to do. I was holding the stick forward by that time to keep the plane from flying and I hadn't pulled the power off yet. So here I was cruising along about 25 feet off the ground and the end of the runway was coming up really fast, because I must have been doing 80 or 90. So I jerked the power off, and when I did, with the forward stick, I just ran right down into the runway. Broke the nose gear clear off,

Fig. 4-17. Small VariEze in flight over the vast desert near Mojave.

spread the gear out—the main gears were spread out so far that the brake discs drug on the ground. And that was the first test flight experience. After I repaired that—it only took five days to repair the nose gear—a friend was at the airport who had flown an Eze, and he flew it around the pattern a couple of times and said it was just fine.

Ed: My first flight was really a super exciting thing for me. I broke ground and the first thing the airplane wanted to do was turn right. So I was busy trying to level off and get some trim into the airplane. It wanted to nose down as I didn't have enough up trim in it. I was holding it from turning right and nosing down with the stick to compensate trying to get trimmed. By the time I got it all trimmed out and looked around, I was climbing 1,000 fpm and I thought to myself, "Holy smokes, I've really got something here!" I brought it around and it took a considerable amount of gumption to try to land it because I thought here I had a dream machine and I didn't want to ding it up. After four or five hours, I got acquainted with the machine and realized I had been flying it too slow for landing.

Don: Most of the people I've talked with had no electrical system in theirs, so they just had a small motorcycle battery. Mine had full electrical instrumentation and an alternator. So I had a full-size aircraft battery right up in the nose and from the very first landing, mine didn't float like everybody else described. I'm convinced that the reason mine handles so docilely in the pattern and comes down so fast is because the airplane has never been smoothly finished; it's never been filled to contour anywhere; the bottom of the wings are extremely rough. They have never even been feather-filled and are very, very dirty. Then with that heavy nose, when you come around the pattern to the end of the runway, all you've got to do is pull the power and you drop 1500 to 2000 fpm; when you get down to the runway, you flare it like a regular airplane and it settles immediately. Mine floats very little and I'm sure it fits the description of how a large biplane lands because it's so dirty that it comes down like a rock—and that's how mine comes down. It would be interesting to see if it would come down differently if the wings were smoothed out and the nose only had half as much weight.

Ed: Personally, I think it has more to do with the weight and balance of the airplane.

Don: It's amazing because so many people have talked about floating and there's no question about it the plane will float. If you drag it in under power, drag it in low and carry power, don't pull it back to, say, 800 rpm, it will float and you'll use up an incredible amount of runway. But if you pull the power all the way back and have the thing idling about 700 to 800 rpm, it comes down right now.

And another point. I talked with people with VariEzes doing snap rolls when they were practicing stalls, and it's obvious to me that the reason for that is that their nose is too light—they simply don't have enough weight up there. My plane has a transformer for strobes in the nose and the battery weighs about 25 pounds, so up there I must have 30 pounds at least in the nose. I think that's why mine does not snap roll. It stalls straight forward; it will do the Dutch roll that's described if you get it down around 70 or so; but I never fly it that slowly. I never fly the airplane under 80 when I touch down. You can rotate with the canard up at 60-65 mph, but it always leaves the ground on its own at 80, and when I touch down it's 80.

Ed: Those numbers are almost exactly the same as my numbers. Mine also stalls straightforward; I don't get what they typically call a Dutch roll.

Don: When they came out with the cuffs for the wings to eliminate the roll, I thought about it for several months. I decided that the Dutch roll had a very effective purpose to it. And that is that in tricky wind conditions or any kind of condition where you're in the pattern trying to follow another airplane or something like that, the Dutch roll has a very, very useful function. It is an extremely effective stall warning. What happens is, and especially with a very low-time pilot, that airplane has no stall warning. You slow it down, you have the power off and you slow it down; as your rate of descent increases, you don't get any sensation of that, or very little. But with the roll that comes in, your airplane is telling you that you're flying too slowly. It's an extremely effective stall warning. You get a controllable situation that whenever you feel any instability, you can dump the nose very fast. I think it saves or can save situations that might otherwise get very sticky. So I have not installed the cuffs, and I have no intention of installing them.

Ed: That's basically the same conclusion that I came to as far as the cuffs were concerned.

What mechanical problems did you have?

Several mechanical problems were mentioned by these three VariEze 500-hour pilots. One had exhaust pipes that constantly cracked, as many as ten pipes total. He even carried a spare stub pipe around for a while before solving the problem. Then there was too much toe-in on the wheels so that tires lasted only 20 or 30 landings. The toe-in was eliminated and new tires now last for 150 landings. One pilot had two blowouts landing in a 30-knot wind with too much rudder and brake at the same time. Side loads popped the tube right out of the tire. Now he always carries a spare tire and tube.

Brake pad problems occurred where the pads would wear out in 150 landings. One pilot's taxiway has a noticeable slope to the right that cause disproportionate brake wear; one pilot had been carrying too much power while taxiing, increasing brake usage.

Les had more than his share of problems that could only happen in a VariEze. As he repaired the minor damage to the nose from a gear-up landing, he put the canard into position to add weight on the nose, but he didn't bolt it into position. Yes, he started his takeoff and the canard came off at about 65 mph. The canard was attached only by the control rod which broke, pulling the canard off sideways so that it cleared the canopy and prop. The canard tumbled down the runway without sustaining structural damage, and Les came to a red-faced stop as members of his flying club watched from the sidelines.

What other unusual incidents do you recall?

Ed: I was losing a lot of oil and I installed an oil separator. I decided the place I should dump the oil back into the engine was via the valve return line on the Lycoming. I installed it and was merrily flying along up over the foothills overlooking this airport that I had never seen before. I looked down and my oil pressure was flickering around 20 pounds. It was kind of a sickening feeling looking down and seeing my oil pressure so low. I decided now was the time to make an emergency landing. An analysis showed that oil was being pumped from the valve return line back up into the breather and it was just blowing it out. When I took off a couple hours before, I had six full quarts of oil—I had just changed the oil—and in that time I had dumped four overboard. So I disconnected my oil breather screen, put some more oil in it and checked the engine. Everything was then okay.

Les: One incident I had was on a long cross-country. I picked up another VariEze builder at Santa Barbara who was flying up with me to Watsonville. To give him room, I put a jacket behind the back seat. We're flying along near King City and the engine quit. I knew I had lots of fuel so I thought I had a terrible fuel leak. The ground was covered with fog except for a mile strip around Greenfield. I tried to make it back to King City Airport and couldn't get through the fog, so I just pulled up and landed on the street there at Greenfield. I started checking things out. The Highway Patrol, three fire units and the Sheriff showed up about that time. It turned out that my jacket pinched off one of the fuel lines, so I had one full tank and one empty tank. I took care of that and made it on into Watsonville without any more incidents.

Ed: What you had done is switched to the aux tank after you "ran out of gas?"

Les: Yes, and made my landing with the engine running—it wasn't a dead stick.

All three Eze builders on the tape mentioned that being the first, or among the first, to land their canard designs at smaller airports always aroused a keen interest from pilots and controllers that added ground time to their travels. They keenly enjoyed the kindness and hospitality of people at these airports.

Much of Don Shupe's flying was in weekly commuting between the Los Angeles area and Reno, Nevada, up over the High Sierras. Shupe reported some discomfort for him and his wife, Bernadette, while flying at 10,000 to 11,000 feet in trying to keep their feet warm. Electric socks solved the problem.

Bernadette Shupe rode the back seat in the VariEze for over 350 hours of Don's flight time. She described her back seat activities in the newsletter for the International VariEze Hospitality Club, which is detailed in Chapter 9, as follows:

"When asked what I do in the back seat on a long flight, I'm not sure people really want to know! I'm sure it depends on who the pilot is, but Donald likes me to take the stick and give him a break from time to time. Other times, I calculate ground speed over the VORs and tell him ETAs along the route. I look for other aircraft a lot. When we start into a new airspace, we both listen to ATIS and Approach Control. Sometimes it's very hard to understand what they want. One night we went into Las Vegas for the first time and we both had a terrible time seeing the runway with all the other lights around. They finally blinked the lights for us; then we saw them. Since I have taken ground school and started with my flying lessons, I am better able to understand the problems and I can be more helpful without getting overly concerned with whether or not what we are doing is all right or safe. I have learned to love it and I miss it as much as Donald does when we don't fly for a while."

Accidents Do Happen

Any new program will develop problems. Flight testing of the many VariEzes is no exception, though the non-spinnable canard design has a far better track record than most homebuilts.

The following accident and near-accident situations were reported to builders in Rutan's *Canard Pusher*.

By far the most hazardous situations occurred when new pilots took off with the VariEze canopy unsecured. Pete Krauss took off without his canopy locked and it opened wide at 100 mph during initial climb. He grabbed it, pulled it closed with his fingers and held it while he returned for a good landing.

Not so routine was the experience of Tony Ebel whose canopy latch was adjusted so loose that it allowed the canopy to rise and fall noticeably in flight. His flight was recorded in the newsletter by Burt.

Tony was flying at 6,000-foot altitude and 185 mph true(165 indicated) when the canopy opened. He doesn't remember if he had bumped the latch. When it opened the airplane immediately departed from controlled flight, yawed, pitched down past vertical, did a ¼ turn spin, then pitched up. Tony grabbed the canopy; it was pulled from his hand and the airplane repeated the above maneuvers. This happened about six times until he finally got the canopy closed with fingers outside (Tony did not have the knob installed on the inside). Once recovered to level flight (only 800-foot altitude), he noticed that his prop was stopped and thus he had to make a forced landing. While the prop will windmill down to 60 knots, once stopped you must go above 120 knots to restart. Tony's engine failed due to negative "G" at a speed below 60 knots during gyrations.

Tony's airplane dug a large hole, cartwheeled once, tore off the right wing and ended up inverted. Tony dug himself out and found that his injuries were minor—cuts and bruises. The airplane was extensively damaged. Since this was the first major overload failure condition on an Eze structure, I was quite interested in inspecting the modes of failure. I flew over the next day and observed the following: wing failure occurred in the spar caps, 3 to 6 inches from the wing fitting—there was no damage in the fittings, winglets failed either in the wing or at the winglet ½ span; the joint did not fail; the canard, itself, was not damaged. All seatbelt fittings were intact. The canopy plexiglass was broken in front, but the canopy frame was not damaged. The forward fuselage back to the instrument panel was totally destroyed. The rear seat area,

Fig. 4-18. Canopy latches can be seen in this close-up of a VariEze with Mike Melvill in the cockpit. Note full shoulder harness for crash restraint.

fuel tanks, c/a spar, fuselage tank, etc., were undamaged. The engine mount, firewall and everything in engine area were intact and not damaged. Nosegear strut and all its fittings were undamaged. The maingear tabs failed. The gear strut failed at ½ span.

"I am confident that inadvertent canopy opening cannot occur if the canopy is built and adjusted properly and locked before takeoff. The handle should be rigged so it must be forced hard forward to engage the latch and handle should be rigged for preload toward each other. Thus it is impossible to open it by bumping the handle. It should take two hands to open. Be sure the latches engage fully in the positions shown on the plans (Fig. 4-18). *Do* install the warning horn that sounds if takeoff is attempted without canopy fully locked. *Do* use your checklist. *Do not* omit the canopy inside knob."

Six newsletters and a year and a half later, an unlatched canopy caused this fatal accident just after takeoff from an airport in Texas. Rutan's report on the accident appeared in the *Canard Pusher*.

"The takeoff and initial climb to a few hundred feet were observed to be normal; then the aircraft appeared to lose control, descending in a very steep angle and crashed about three miles from the airport. Both occupants died instantly. The aircraft had over 80 hours flying time. It had been tested extensively and had never demonstrated any unusual flight characteristics, according to the previous owner. Working with the FAA investigator, we found no indication of structural failure, control disconnect or engine failure. An examination of the canopy was not locked at impact. The aircraft was equipped with the canopy safety latch and its damage showed the canopy was open approximately 1 ½" and engaged in the safety latch at impact. There was no canopy unlatched (light/horn) warning system nor inside canopy closing handle installed. The aircraft was within the allowable gross weight and slightly aft of the aft cg limit.

"Since the canopy was unlocked at impact, it would indicate the pilot had failed to complete his take-off checklist and took off with an unlocked canopy. It appears that the distraction of the canopy opening against the safety latch, combined with a possible panic-stricken passenger (it was the passenger's very first airplane ride) might have caused the pilot to lose control of the aircraft.

"There are three other cases of VariEze canopies opening in flight at this time. All three were able to control the airplanes to a landing, even though they were holding it down with fingers outside the canopy frame (two inches open). With canopy open (2″) in the safety latch, other than a moderate wind blast, VariEzes can be controlled and landed normally. The following is a first-hand account of such an incident by Les Faus:

"Burt asked me for a few words about how it is to fly a VariEze with the canopy open. Mine opened at about 50 feet and 100 mph.

'I fortunately had a back seat passenger that I could rely on. Between the two of us we were able to close the canopy without too much trouble. With the canopy full open, the plane tends to pitch up and to the right. I put the stick into the left front corner and eased back and just held it straight until we could ascertain the damage. The back seat passenger held the canopy closed while we flew 15 miles to another larger airport for landing. The airplane flies well with the canopy being held open with your hand around the frame. (About 2″ open.) At that time there was no safety lock on mine. The only damage to the canopy was the center arrow stock broken by the back seat passenger trying to close the canopy. If this happens to any of you, don't panic! The airplane is controllable and can be saved. It sure gets the hair up the back of your neck at the time, though!'

"Slow-speed approaches have given VariEze pilots some problems. During a winter landing in Michigan, one pilot was slow on final, developed an excessive rate of sink, hit hard, then cartwheeled while bouncing about 30 feet into the air, and coming to rest inverted in a snow back. Other than a small abrasion on one knee, the pilot was not hurt. The Eze was substantially damaged. The pilot had 2.5 hours in his VariEze, all operating from a large airport. He had not flown in the previous three weeks due to bad weather. On this flight, he was attempting to land on a 2850-foot runway with powerlines and trees on both ends in a ten-knot crosswind. The runway was covered with snow except for a 45-foot wide path in the center and there were 3.5-foot-high snow banks on each side. Ground witnesses reported the pilot attempted two landings but got slow on each and went around. On the third approach, he once again was really slow in a nose-high (estimate 30°) attitude, developed heavy wing rock and a high sink rate. He hit hard, spread the main gear strut straight out, then caught the snowbank and cartwheeled."

However, the track record of the VariEze design has been very good when compared with other homebuilts. In the first 193 to be registered, FAA statistics show that only five were involved in accidents, three of which were fatal. This figures out at 1.98 percent of the Rutan designs involved in accidents, compared with an industry average of 3.93 percent for all amateur-built aircraft.

THOSE UNFORGETTABLE FIRST FLIGHTS

Rich Clark of Hermosa Beach, California, submitted the following account of his first flight in E-Z-GO:

"Liftoff surprised me. Continuing acceleration with the stick slightly back is a lot different from stabilizing speed with the nose held down. I tried mentally to freeze my hand and succeeded in a series of damped pitch oscillations during initial climb. Roll/yaw okay. I let out a Texas yell, wiggled the wings for the congregation and settled down to trim out. Descent and approach were smooth. Love those ailerons. Over the numbers at 80 knots, overflared up to 20 feet, and another series of pitch oscillations. Eventually, E-Z-GO got tired of the comedy and sank to the runway. The solid rumbling of gear on runway and straight rollout were reassuring. Why could I not drive it on? Main factors were pitch sensitivity and Cessna training. Also, confusion with left-hand throttle where up is go and down is slow.

"Later—feeling at home expanding the envelope. Sure enough, won't stall, just nods. Heavy back pressure at low speeds. Easy to locate traffic up here in the bubble. High on final. I ease back to 70 knots to below glide slope. Ease power on to hold it. Good flare and soft rumble and I'm at jogging speed. Had a lot of confidence before I had flown it; now I have much more."

Sometimes first flights are not always reported from the cockpit. Mrs. Helma Hamel described watching her husband Ellsworth's first flight in VariEze N235 EH this way. "It was a tremendously scary and exhilarating moment when Ells started climbing. There is nothing in my life that can compare to the shudder and thrill I felt—not even the birth of our first child!"

First flights with Dane and Rudi's HB-YBG were made by Gion Bezzola, a Swiss Air Force instructor and jet pilot. He detailed the flight program for readers of the newsletter.

"After putting the plane together and carrying out a comprehensive pre-flight check, according to the American Handbook, I did two taxi tests with the nose up, and another with about 30 percent power to test the elevators and to get used to them. The nose allowed itself to be lifted from the ground at about 50 mph, and to be held there. The VariEze began to dance about on its toes as if she couldn't wait to get into the air. I wished to remain captain in charge, so we stayed on the ground a while longer to test the steering. I had to remember to bring my feet back when I wasn't using them; otherwise, I would have inadvertently used the brakes. The acceleration was very good. We were very excited before the first lift-off, but this was carried out with no cause for alarm. I got used to the lateral position of the stick very quickly, and then the Eze was flying, one meter above the ground down the runway. I held the speed at about 80 mph. I decelerated by throttling down slowly, lowered the plane and landed like a feather. The steering felt finely balanced and the brakes were adequate. After four more short flights at a height of 1-2 metres, the inspectors from the Air Ministry arrived and then I was ready for the first big flight of HB-YBG.

"The five short flights, the taxi tests, the performance of the engine, and the faultless work that the builder had done gave me a lot of confidence in the machine. We worked out that the center of gravity lay in the allowed area when I was wearing a parachute. The motor was running quietly, and so I asked for starting permission from the tower. That moment was here again—the first big flight in a new plane—a fantastic exciting moment. I had already experienced such a moment when I test flew my own construction, the Lutibus, HB-YAY, and now I was looking forward to it again. I opened the throttle, lifted the nose at 50 mph; at 70 mph I lifted the Eze from the runway and held the resulting angle of climb. The angle of climb came as no great surprise to me; as a jet-fighter pilot, I am already used to such a steep angle of climb! But the fact that this was a homebuilt machine powered by only 90 hp—that was a fantastic surprise.

"We, the Eze and I, had hardly started, and already I was crazy about her. The steering reacted marvelously, and I trimmed it out with short blips on the electric trim switches. I had to keep watching out that I didn't exceed the permitted maximum speed for a wheel-out flight, because the plane kept wanting to go faster. The Eze was a real little thoroughbred. The air-brake worked very well when it was out and showed this by slightly noticeable buffeting.

"I realized something pretty quickly; this was probably the most phenomenal plane that I had flown, and I had flown forty different types, Mirage and Hunter included.

"After about twenty minutes, I carried out a simulated touch and go and could judge my imagined landing point very well, a fact that was later supported by the actual landing. On final, at about 90 mph, then over the beginning of the runway without about 80 mph on the clock, the Eze was landed with enough pilot visibility. I landed the Eze and held the nose up in order to brake aerodynamically. Rolling down the long runway gave me time to savor the thrill of that first flight. I was also pleased for the builder of HB-YBG, who had, through his extremely clean work, given me one of the best flying experiences of my life.

"The loud 'hellos' from the spectators who had gathered in the meantime, the happy smile of the builder, Rudi Kurth, the congratulations from friends and from the officials of the Air Ministry, were payment enough for the preparation made for this flight, that was threatened at no time by an uncalculated risk. The knowledge that the VariEze designer, Burt Rutan, knew exactly what he was recommending to future Eze pilots through the handbook, proved to me that a very conscientious pilot was the spiritual father of a brand new type of plane. This knowledge grew stronger as the days passed, especially when I took the Eze through the stall test and through the largest part of the tests. How many crashed planes and their pilots could have been saved had this type of plane been designed earlier, because the behaviour of the Eze in extreme conditions is simply fantastic. That a plane can still fly controlled turns when it is stalled, and climbs with full throttle without the slightest danger of a spin is a wonderful performance of modern aerodynamics.

"The manufacturers of conventional planes will have to think again if they wish to equal this type of safe flying. The influence which the VariEze will have on general aviation cannot yet be judged, but I have the feeling that it will be a great influence. After the first flights, I was sure that Rutan had not promised too much. The plane which Rudi Kurth had built was exactly according to the specifications which could be found in the handbook."

Chapter 5
Canard Concept
and Composite Construction

Rutan's basic configuration—whether the VariViggen, VariEze, Long-EZ or Defiant (Fig. 5-1)—is a loaded canard wing designed to provide aerodynamic angle-of-attack limiting and thus be stall-proof.

When the opportunity comes along (as it did to us on several occasions) to fly Rutan's designs, you'll really enjoy this proof of concept in the air.

THAT LITTLE WING UP FRONT

The Rutan crew describes the canard system very simply. A canard airplane has a tandem wing—one wing in front of the other (Fig. 5-2). Generally, the front wing is smaller than the other. Mike Melvill commented that he knew of no canard where the front wing wasn't smaller. The front wing provides pitch control, though the Swedish Viggen's pitch control comes from elevators on the back of the delta wing. Dick Rutan says that this makes it not a canard, while Burt maintains that it is. So you take your choice. The Eze canard design has flaps for landing on the back wing—not elevators.

Both wings on the canard design are lifting rather than having the tail of a "conventional" design with a down load that causes drag. Nothing on the canard design is "lifting down."

The canard, itself, goes back to the days of the Wright Flyer. Canard glider designs included Wolfgang Klemperer's in the 1920's that was supposed to sense rising air currents and pull up into them to gain altitude. In WW II, Curtiss-Wright built a canard intercep-

Fig. 5-1. Rutan's canards in formation. The twin-engine Defiant, the VariViggen and the Long-EZ in the desert skies at Mojave.

Fig. 5-2. Burt Rutan in the twin-engine Defiant, Mike Melvill in the VariViggen and Dick Rutan in the Long-EZ.

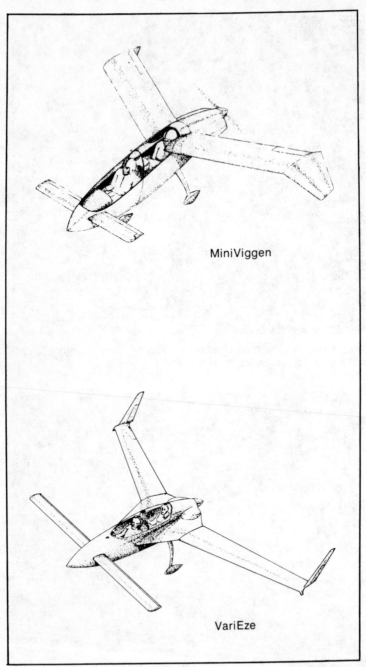

MiniViggen

VariEze

Fig. 5-3. Initial sketch of the "MiniViggen" and an early sketch of the VariEze.

142

tor fighter with a pusher Allison engine called the "Ascender," which was the subject of numerous snide puns.

After Rutan's early success with the complex-to-build VariViggen, he turned to a smaller, cheaper, hopefully less-complex canard. His initial effort in this smaller class of airplane was called the "MiniViggen (Fig. 5-3.)" As Burt reported in an article carried in the January, 1976 issue of *Sport Aviation*.

"It was a high-wing, low-canard, two-place with fixed gear and conventional ailerons. Its structure was all aluminum, using forward fuselage formed skins and canopy from another design and single curvature aluminum sheets for the aft fuselage and flying surfaces. The aircraft was designed around the then-available 60-hp Franklin aircraft engine. In early 1974, I built the fuselage and canard of the MiniViggen. I later scrapped the structure because I found that the compound curves of the fuselage made the fabrication and installation of all internal parts very difficult and time-consuming. I was also not satisfied with the lack of insulation, vibration, points of fatigue, and weight of the metal structure. I later found through model testing that the design had negative

Fig. 5-4. Winglets are to be found on the latest Gates Learjet Longhorn 28/29, the first production aircraft to use this advanced winglet technology. Rutan has been flying winglets on his homebuilts since 1975 (courtesy Gates Learjet).

Fig. 5-5. The company's most recent model, the larger Learjet Longhorn 50 series is in the foreground accompanied by an early model 23. The 50 series has widebody comfort, carries 10 passengers, flies well over 3,000 miles and at altitudes to 51,000 feet. The original 23 had a relatively small cabin, carried six passengers, had a maximum range of approximately 1,400 miles and was certified to a maximum altitude of 41,000 feet. Winglets are used to increase efficiency on the new 50 series (courtesy Gates Learjet).

spiral stability and a region of pitch instability at the approach speed.

"During October/November, 1974, I learned of NASA's yet unpublished research with Whitcomb winglets (Fig. 5-4, 5-5), contacted Dr. Whitcomb for details, and incorporated them into the design. I had been studying the smooth contoured, efficient glass composite European sailplanes. These all-glass ships represented to me the only really significant advance in lightplane aerodynamics and structures since the advent of the Beech Bonanza—it's a shame they require expensive female molds, exotic materials and skilled craftsmen to build, for a homebuilt composite aircraft would sure be nice with its light weight, wrinkle-free structure, improved corrosion resistance, longer fatigue life, and a dramatic reduction in the number of parts!

"In order to demonstrate the efficiency of the configuration, I decided to develop an airplane to capture Ed Lesher's speed and distance records in the under 500-kg weight class. I spent the next two months in detail design of the VariEze prototype, using a simple square-inside box fuselage and solid core flying surfaces. Having no previous fiberglass experience, I built N7EZ (Fig. 5-6) over the next 3½ months. The first half of the construction project was done alone, with occasional help from Carolyn. During the second half, Gary Morris donated his evenings and weekends to help. Phil Rathbun built the machined parts for us.

"N7EZ was assembled at the hangar for initial tests on May 21, 1975. That day it made several high-speed taxi runs and one short 10-foot altitude hop down the runway. It was not a good day; nose gear shimmy required repair and redesign; nose gear rotation speed was too high and roll control inadequate due to stalled elevons. That night a close documentation of the aircraft was done and it was found that due to a variety of errors, the canard incidence was too low with respect to the wing and the wing had only about half the amount of washout called for in the design. A temporary strip of sheet metal was taped to the wing trailing edge and bent upward to reduce the wing's pitching moment (this was later removed and the wing's under camber filled to compensate). The canard's incidence was also increased, and the next day the Vari-Eze was off on its first real flight. After some fine tuning to get the proper canard incidence, the airplane was pronounced satisfactory for formation flying and was tucked in under the photo chase required for *Air Progress* magazine four days after first flight.

"Analysis and tests indicated that a very poor maximum lift coefficient and early airflow separation existed on the canard

Fig. 5-6. Burt sits at the controls of the original VariEze N7EZ as his mother Irene inspects the back seat for space. She was later to fly in this prototype (courtesy Burt Rutan).

causing the high stall speed (60 knots) and poor roll control below 80 knots. The 14-inch chord canard was operating at a very low Reynolds number and the GAW-1 airfoil could not exceed eight degrees angle of attack and 19 degrees elevon position without at least partial stall. A little reference searching and the help of Dick Eldridge of NASA's Flight Research Center resulted in finding an airfoil designed to operate at low Reynolds number—the GU25-5(11)8 developed at the University of Glasgow. The GU25 offered several advantages:

☐ 50 percent greater G_L than the GAW-1 at Reynolds number of ½-million

☐ thicker section resulting in a lighter, stiffer canard

☐ no trailing edge camber, thus eliminated the requirement for external trim tabs

☐ attached airflow at higher elevon deflection.

"I dusted off the cartop windtunnel and tested the two airfoils to verify the published data and to evaluate three different elevon gap configurations. The new canard was then built and flown. The difference in the airplane's flying qualities was the most dramatic change I have seen without a planform change. Whereas before, the aircraft had a nose-down stall break at 60 knots, it could now be flown at 52 knots with full aft stick! Whereas before, the roll rate was poor below 80 knots and had to be supplemented with rudder below 70 knots, it now had satisfactory roll control down to the stall speed. Roll rate at all speeds is now much better than a Cessna, but less than a Yankee. According to windtunnel data, the cruise speed should have been reduced three knots; however, flight tests showed identical cruise performance and in improvement in low-speed performance (maximum L/D increased to 18.5)."

Rutan expanded on the desirable features of the canard design for use:

"Regarding pitch stability and high angle-of-attack characteristics, the canard arrangement obviously does not guarantee stability and a safe stall—rather it gives the designer the flexibility to tailor the characteristics to his desires. What is desired? The optimum is a linear airplane in the normal flight regime with a strong stable break in the pitching moment curve near maximum usable lift. The conventional one-wing, tail-aft configuration does not allow a designer to provide natural limiting of angle-of-attack. If elevator power is strong enough to get the nose up at forward cg and is capable of driving the main wing beyond initial stall to where loss of directional stability or massive unsymmetrical wing stall

Fig. 5-7. Canards of all sizes and shapes are crowded into RAF hangar at Mojave. In the foreground is the VariViggen, at the right the nose of the Defiant, then the prototype VariEze and the Long-EZ in the background.

can occur, these, of course, are the causes of spin susceptibility. There is no need or desire to ever operate beyond the initial stall. Any student knows that yanking the stick further aft at the stall will result in you being a pile of mush on the runway. But if the airplane limits itself at the initial stall and has no stall break, then we have real stall safety. The nose on the Defiant does not "drop" at the stall. It stays at its maximum angle and if the cg is aft, it gently "nods" about one degree every three seconds. The pilot retains full control of flight path at this extreme condition of holding full aft stick. Even with a failed engine, he can obtain an instant climb, or descent, or fly indefinitely level while holding full aft stick. By contrast, a conventional aircraft, if stalled, will mush to the ground even at full power, since the elevator power is sufficient to cause massive wing stall with its attendant loss of lift and enormous drag (reference stall/mush accidents on takeoff and landing).

"Why can the canard (Fig. 5-7). do this at forward and aft cg? The reason at forward cg is simple, since the definition of the forward cg limit is that cg beyond which the low-speed performance is not obtained. Now, let's see how the canard configuration can be designed to limit angle-of-attack at aft cg. Elevator power at aft cg is considerably higher, and the high angle-of-attack is reached with very little elevator. Then, as the canard begins to stall (a gradual loss of lift, not abrupt, if the correct airfoil is used) a great deal more aft stick is required to get only a small amount more angle-of-attack. At aft cg using full aft stick, the canard is actually operating above its maximum lift angle; thus the canard is a stabilizing surface (more angle—less lift), and pitch stability is much 'stiffer' than in normal flight. Thus, at this flight condition, the aircraft will maintain angle-of-attack very accurately. Contrast this with the normally 'sloppy' pitch stability of a conventional aircraft at the stall, and its abrupt loss of lift."

In speaking specifically of the twin-engine, push-pull Defiant, Rutan explained: "The main reason the Defiant out-performs the other light twins is its low wetted area, not its canard configuration. However, the reason it has low wetted area is because of the canard configuration."

"The canard configuration drastically reduces the magnitude of the structural loads and eliminates a very large percentage of wasted wetted area. Its simplified design philosophy eliminates many unneeded systems like flaps, cowl flaps, etc. Its control runs are shorter and carry less than half the forces. All the above items contribute to weight savings, which reduces required wing area, again saving weight. The next result is that even though the Defiant's ultimate load factor is 50 percent higher than the conventional twins, it is nearly 1000 pounds lighter and has less than 2/3 the wetted area. Again, the reason is the canard configuration."

1990 DESIGNS AND CAPABILITIES

Rutan is by no means the only aircraft executive who is high on the use of new construction techniques and material and the canard concept. Malcolm S. Harned, Senior Vice President, Technology, Cessna Aircraft Co., has some very definite ideas on the subject. Some of these were expressed to the American Institute of Aeronautics and Astronautics (AIAA) at a seminar in Washington, D.C. under the title, "General Aviation Aircraft—A Forecast of 1990 Design and Capabilities."

"Several factors will assure a high demand for general aviation aircraft through the decade of the 1990s, namely, increasing airline specialization in mass transport between major hubs and, as a result, greater use of private or executive-type aircraft by businessmen and the affluent for both convenience and comfort.

"This trend will foster a need for more feederline aircraft. There also will be broader demand for increased safety, fuel efficiency, comfort, performance per dollar as well as reduced maintenance.

148

"New technologies will be available such as composite materials, new aerodynamics, very sophisticated electronics, fallout from the auto industry's booming technology, and advances in turbo-machinery. New generations of aircraft will materialize, most with pressurization, all-weather capability, and self-monitoring diagnostic systems to minimize failures and maintenance requirements.

"Reduced weight, improved aerodynamics and engines will generally increase speeds by 25 per cent and kilometers per liter by more than 50 per cent. Compound aircraft will provide vertical landing plus high-speed cruise. Efficient short-haul feederliners will also be available.

"The airlines will become extremely efficient transporters of masses of people for long distances. As a result of the recent reductions in air fares here, there were large increases in traffic.

"However, worse than the crowded aeroplanes are the congested terminals. These two factors make flying very unpleasant for the businessman and eliminate the possibility of working while travelling.

"In addition, there is a continuing decline in airline service to smaller communities. Since the quadrupling of the cost of fuel, the airlines can no longer afford to service their low load-factor routes, which generally are to the decentralized business and industry communities.

"In the smaller communities there are rapidly growing numbers of people who want to take advantage of the low-cost, high speed air travel available at the hub terminals. This will create a very large demand for feederline operations. These aircraft need to be relatively small, from 9 to 50 passengers, to provide reasonable frequency of service and be economical to operate to the smaller communities.

"Fortunately, there will be a number of new tools to assist the aircraft industry in meeting the stringent demands of the greatly expanded marketplace.

"Great promise for general aviation is offered by composite materials, such as the aramid fibers and graphite fibers with an epoxy bond. Both types of fibers offer strength-to-weight ratios and modulus of elasticity-to-weight ratios very superior to aluminum alloys.

"Before this potential is fully realized, there are many developments required which include lightning protection, inspection and testing techniques, interfacing with metals, new approaches to structural analysis and design, new manufacturing techniques and methods for field repair. In addition, material costs must be drastically reduced.

"However, by the 1990s these problems should be resolved and these materials should be standard production items. The fact that Kevlar® is replacing steel in premium tires today is very promising. They will be used not only for basic airframe structure but for propeller blades, landing gear, etc., with the general result of at least a 25 per cent reduction in empty weight.

"There should also be significant improvement in aerodynamic efficiency as the result of the universal application of refined versions of the so-called supercritical airfoil sections. These will not only be applied to wings but also to improve the efficiency of propellers.

"Aircraft piston engines will be significantly better, both in power/weight ratios and specific fuel consumption. Composite materials will be used extensively for engine structure and components, thereby reducing weight. Lean burning techniques with fuel injection and other improvements should also offer 10 to 15 per cent reductions in specific fuel consumption (SFC). Even diesels will become usable with over a 25-per cent improvement in SFC.

"The use of pusher propellers will be made practical as a result of using composites for lightweight, very reliable drive shafts and gear boxes.

"This approach offers several advantages:

☐ The high speed propeller slipstream does not impinge on the aircraft, thus reducing drag.

☐ Mounted on the tail, the inflow to the propeller keeps the air flow attached to the tailcone, also reducing drag.

☐ Nacelle drag can be eliminated by locating the engines in the fuselage.

☐ The aft location substantially reduces cabin noise.

☐ Visibility is greatly improved.

☐ For twins, gearing the two engines to a single rear prop gives center-line thrust, eliminating any yaw with an engine failure. This configuration is also ideal for thrust reversing.

☐ The rear location is much safer on the ground since with the proper tail configuration the possibility of people walking into a propeller can be virtually eliminated.

"The twins will all have thrust on or near enough the center-line to eliminate V_{mc} (minimum control speed for twins) as a consideration which will be a real safety advantage.

"Most [aircraft] will have engine monitoring systems which will sense vibrations, torsional loading and metal in the oil to anticipate engine failures well in advance. This will not only increase safety but will reduce engine maintenance costs.

"The six-place and larger aircraft will have strain gauge systems mounted on the landing gear that will make possible an automatic weight and balance readout from the computer.

"The 25 per cent lower empty weight, the supercritical airfoil and the full-span flaps all combine to make possible reducing the wing area by one-third. However, the wing span has been retained to give good climb characteristics with relatively low power and a high lift/drag ratio for the higher altitude cruise.

"The minimum family twin is illustrated. To provide minimum cost, two turbocharged automotive Wankel engines are used. Their compact size and light weight almost make possible a convenient arrangement for a center-line thrust twin to provide maximum safety. Since these engines are liquid cooled, the radiators will be aluminum leading edges on the wing for the front engine and on the tail surfaces and inlet for the rear engines. This will also provide an automatic anti-icing capability.

"Although the Wankel will always be inferior to the piston engine in SFC, its light weight, compact size and lack of vibration will perpetuate its development as an automotive engine with the result that its low cost could make it attractive for personal aircraft. The lack of a valve train and basic simplicity should make it very reliable.

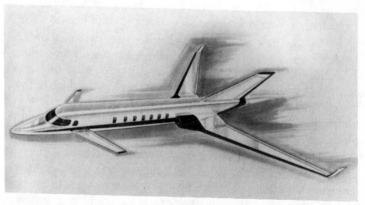

Fig. 5-8. Mach 0.95 business jet of the 1990's as foreseen by Malcolm S. Harned. This artist's diagram shows highly swept wings with super critical airfoils and a canard surface on the nose. Looks a lot like Rutan's designs today (courtesy Cessna Aircraft Company).

Fig. 5-9. The latest Learjet, closest to the camera, features winglets. Buried engines suggested by Malcolm S. Harned are still in the future. Bottom to top of photo, Learjet Longhorn 50, Century III 35A, Century III 25D and early model Learjet 23 (courtesy Gates Learjet).

"Continuing up the scale in speed, in that decade we'll see the Mach 0.95 business jet (Fig. 5-8). This would offer essentially a 20 per cent increase in speed over today's business jets and at the same time provide high fuel efficiency.

"It will be necessary to bury the engines, area rule the fuselage, go to highly swept wings with super-critical airfoils, and a canard surface on the nose to minimize trim drag. Winglets (Fig. 5-9) will serve a dual purpose, increasing the aspect ratio and directional stabilization.

"This aircraft would offer stand-up aisle height, 16 places plus a 965 km/h cruising speed capability at altitudes up to 60,000 feet with ocean-crossing range. Even at this speed, it should offer a fuel efficiency of 1.7 km per liter. It also offers the safety advantage of essentially center-line thrust plus a cabin free of engine noise.

"Another new category for general aviation will be the short-haul commuter transport. Although there has been a limited participation in this field with aircraft derived from business airplanes, this market will grow in size by several times in the next 15 to 20 years. Consequently, there will be all new designs developed in which the principal emphasis will be on the minimum amount of aircraft weight per passenger lifted into the air.

"One approach to such a 50-passenger short haul airliner is illustrated (Fig. 5-10). By using a tandem wing configuration, minimum trim drag is achieved with good control power for low take-off and approach speeds. It also makes possible an aft location of the turboprops, which will provide a very low cabin noise level. The ability to use the aft pressure bulkhead as the carrythrough structure for the main wing minimizes weight.

"It would be pressurized to cruise at 25,000′ where it would achieve speeds up to 480 km/h. Even for relatively short routes, it would offer over 42 seat-km per liter.

"A principal requirement for future general aviation aircraft will be for improved safety. Consider, for example, general aviation's record with those for other modes of transportation (fatality rates per 100 million passenger-kilometers, based on U. S. figures only):

Overall airline aircraft	0.025
Overall general aviation aircraft	10.0
Overall passenger cars	0.875
Passenger cars on turnpikes	0.44
Cessna aircraft types:	
Skyhawks	4.375
421s	0.94
Citations	0.25

"In summary, in the decade of the 1990s, we should expect our general aviation aircraft to generally provide 25 per cent more speed with 50 to 100 per cent better fuel efficiency plus greatly improved safety, reliability, convenience and comfort. They should reduce the accident rate and be safer than cars. These new aircraft, in combination with the expanded marketplace, should assure that the decade of the '90s will see more than a five-fold increase in general aviation business."

RUTAN'S COMPOSITE CONCEPT

Many of Malcolm S. Harned's visions of the 1990s are already being assembled and flown by Burt Rutan and his army of homebuilders.

Burt emphasizes that he has developed *methods* to simplify fabrication, not *materials*. The materials are those proven through years of aircraft experience.

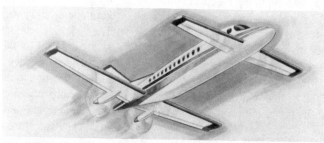

Fig. 5-10. Canard design with pusher props is visualized for 1990 by Malcom S. Harned of Cessna Aircraft Company (courtesy Cessna Aircraft Company).

Burt Rutan gives much of the credit for his new construction procedures to Fred Jiran who operates a glider repair facility on the Mojave Airport. Burt spent many hours in Fred's shop "drooling over the smooth, contoured, efficiency glass composite European sailplanes."

In his initial detailed EAA announcement of the VariEze program Rutan wrote the following (originally published in *Sport Aviation Magazine*), explaining his reasoning behind the composite structure:

A composite as used in this article is defined as a sandwich of a low density core covered on both sides by a high strength material. A major point I want to clarify is that the VariEze is *not* a fiberglass aircraft—it is a *composite* aircraft. The only areas which could be called fiberglass structure are the cowling and wheel pants. Fiberglass used by itself results in a heavy structure that can flex, deform and vibrate and has no real advantage over conventional metal structure. Please do not equate glass composite to any experience you have had with fiberglass cowlings, boats, cars, etc. The next time you see a VariEze, pound on the cowling and then the wing with your knuckles—the difference in sound between the dull "thud" and the solid "knock" will give you an idea of the advantage composites offer.

A review of the many reports of military aircraft and commercial applications of composites over the last few years in *Aviation Week and Space Technology* magazine will show that problems of fatigue and corrosion in metals are a major reason for adopting composites. This advantage is not well known in general aviation circles, however, since I have received many questions from people who were concerned about the life of the structure. Without getting too technical, I will attempt to show the reason composites are superior. To do so, it will help to compare the VariEze structure to contemporary metal design.

A typical aluminum wing is a combination of a spar, ribs and skin attached in several hundred places by rivets. It is common practice to use safety factors of only 1.5, such that the metal permanently deforms at loads lightly over limit, or the maximum allowable flight loads. Further, the typical metal spar or rib is designed without full redundancy such that one crack can propagate across the piece and result in catastrophic failure. Commercial and military aircraft designed in this way require strict quality control and expensive equipment to inspect for hidden flaws and have a calculated fatigue life. The homebuilder generally uses only the visual surface finish to determine flaws, makes periodic inspections, and takes his chances. All flight loads are transmitted in concentration, not only at specific places (ribs), but at discrete rivets. This results in many minute high stress areas and the majority of the structural weight operating at very low stress. Further, if the ribs do not fit perfectly, they are pulled into position when the skin is installed, resulting in "assembly stresses," or loads, not due to flight loads. The metal skins and ribs also wrinkle and buckle under flight loads. Aluminum under stress has a definite life, beyond which it will crack and fail. To have a long life aluminum structure, the designer must assure that all flight stresses are small, and the builder must assure that flaws do not exist due to tight bends, nicks, assembly fit, or improper fastener installation. Due to constraints of complexity and weight, the designer can use redundancy in only critical areas and trust that adequate quality control is exercised when building parts.

In the certification of general aviation aircraft, FAA now requires the manufacturer to calculate the expected average fatigue life of the structure in flight hours and divide that value by eight. The result is the expected life before failure of the worst-case airplane built. You may be surprised to find how low this number is on several of the metal designs now offered to the homebuilder.

In contrast, the VariEze wing has no ribs, no concentrations of high stress, and a spar which is multiply redundant. It is also designed for safety factors of approximately four, instead of 1.5. There is absolutely no wrinkling or buckling of any component, even above three times the design load factor for the aircraft. With the aircraft operating in its normal envelope, the maximum stresses are only a small fraction of the percentage of allowable stress, which means an exceptionally long fatigue life. Further, cracks cannot propagate across individual glass layers or even to adjacent fibers of a layer. Delaminating stresses are low, but even if a major amount of delamination occurred, the wing would still maintain its integrity for all normal flight loads. The nil-absorbent epoxy used is not susceptible to water absorption and freezing which causes crazing on some fiberglass products. Skin durability is such that you can walk on any portion of the wing with hardsoled shoes and cause no damage.

European glass sailplanes have composite wings which are built in two major parts—top and bottom skin—each skin is a composite with glass outside and inside a ½-inch thick foam core.

These parts are layed up into molds and, when cured, are joined over a glass spar, bonded at the spar and wing leading and trailing edge. For the homebuilder, this has several disadvantages: requirement for female molds, relatively high stress carried in the foam, difficulty of inspecting spar bond, and careful preparation of surfaces is required to assure a reliable bond. On the VariEze, I made the entire wing a single composite by using a solid core of low density foam. The homebuilder makes the cores by running a hot wire around rib templates. This method is quick and accurate, and it is easy to obtain any desired airfoil contour and wing twist. Where the box spar is located, the cores are cut with hot wire, the glass shear web is layed up, and then the cores are assembled. The glass spar cap and glass skin are then applied in one horizontal lay-up.

The advantages of this method are many: far fewer man-hours and skill are required, the box spar and the skin are 100 percent foam-supported, all the structure is easily inspected from the outside, and thermal stresses are eliminated since all structure is at the surface and expands uniformly. Foam stresses are very low and not concentrated. There are no stresses that tend to cause the foam to separate. The outer skin is quite stiff compared to the foam; even the foam's normal 2 percent dimensional instability with heat and time cannot give it undue stress. The foam is really just along for the ride, providing local buckling support and cannot change the shape of the wing once the glass skin has cured. The foam's role of providing buckling support is also not very critical to the integrity of the wing. Due to the high safety factors, a large percentage of the foam could be removed or detached from the skin, and the structure would still be adequate.

Do not confuse this construction with the method which uses a wood spar, foam ribs, and non-composite foam skin covered with Dynel. The success of that method is actually a good indication of the safety margin of glass composite, since it apparently does result in a satisfactory structure while breaking most of the rules! First, it is the best candidate for thermal stress; the wood spar is insulated by foam, so when the temperature changes, the skin immediately goes to the new temperature causing thermal stress. Second, bare foam is used in individual ribs to carry lift loads back to the spar, and the skin is foam sheets with one side bare. Thus, the foam is stressed many times greater than with a composite, and foam dimensional changes can affect structural shape. Third, Dynel has a very low modulus of elasticity, so that when stressed, the epoxy will actually crack before the material itself fails. The low modulus means that when the wing is loaded, more stress is dumped into the foam. Dynel is also quite thirsty for epoxy (its weave resembles burlap) resulting in a high weight/strength ratio layup. Dynel's main good point is the relative ease in which it lays down. Dynel or polypropylene *cannot under any circumstances* be considered for a structural application in a composite similar to a VariEze. Graphite or Kevlar® could be used in place of glass cloth, resulting in a 10 percent to 15 percent weight reduction, but the cost is prohibitive.

The glass spar is more efficient than an optimum metal spar, because the cap is uniformly tapered (using scissors when cutting glass cloth, as opposed to a milling machine for aluminum!), the direction of stress is oriented along the load path (along the spar cap at 34° in the shear web), and complete buckling support raises the strength of the compression cap to near that of the tension cap. A hand-layup of undirectional fiberglass with the materials the homebuilder will use in VariEze construction has almost the same ultimate tensile strength, more compressive strength, and is 2/3 the weight of 2024 T-3 aluminum! Since the stresses are more uniformly applied, the strength-to-weight ratio of the entire structure is much greater.

Aluminum is susceptible to corrosion, requiring particular protection in humid, salty climates. Wood is susceptible to dry rot if not carefully protected. The glass structure requires no corrosion protection and can even withstand a salt water environment. The composite cannot trap moisture and has no hidden joints susceptible to stress corrosion.

Every homebuilder and every FAA inspector already has the inspection equipment required to check for flaws—his eyes. Visual inspection for voids or dry areas is all that is required. Due to the available adequate inspection method, ease of producing a quality part without high skill, and high safety factors, I believe that glass composite will provide a marked improvement in structural reliability over aluminum, steel or wood.

While I personally found the hand-layups easy and feel that the main reason is the special weave cloth, I wanted to get an opinion from others. So, I invited a couple of EAA members in to help build a wing. The following are their comments on the ease of working the glass materials. The first has built and rebuilt a metal airplane and has built a fiberglass airplane structure using common industrial glass cloth and polyester resin. The second one has built a four-place metal aircraft and has built some fiberglass parts for it.

Dear Burt

I'd like to take this opportunity to thank you for allowing me to participate in the construction of the wing for the VariEze prototype. I'd also like to make some observations, for the record, on my thoughts about the technique.

Having built an all metal airplane, I was amazed at the difference in the times involved to build the wing. I was conditioned to the literally thousands of operations required to rivet an aluminum wing together and seeing you accomplish the same end result by simply wiping the resin into the foam and glass with a squeegee is nothing short of fantastic!

I had done some foam and fiberglass work on another project and was particularly impressed with the resin and cloth combination which you are using. It lays up smoothly and easily with a minimum tendency to "air bubble." I'd like to use some of the same if I ever get around to doing another foam and glass project, which might be soon.

If you ever need a testimonial for the technique, please call on me. Thanks for the enlightenment.

Sincerely,
Jerry L. Kibler

"Burt,

Having built my own all metal airplane, and helped build two others, and having had a little experience with fiberglass, all bad, I was amazed at how easily and quickly foam and glass could be turned into an airplane. When offered the chance to help lay up one wing for the prototype VariEze, I was at first reluctant, but found the special glass and resin easier to work than any other I have seen. It squeegeed out beautifully, with no tendency to stretch or trap air, and the entire wing was finished in less time than it took to build the vertical fin for my plane!

Steven F. Guilford
Anaheim, California"

A composite structure is a quieter structure; there is no buckling or oilcanning, the foam core is a natural sound deadener. The thermal insulation qualities of the composite are beneficial in the cockpit. Whereas, in a metal airplane the radiant warmth from the canopy quickly escapes through the skin, the VariEze's canopy will even keep the pilot's feet warm at high altitudes without a heater.

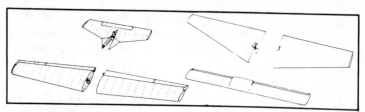

Fig. 5-11. Detail drawing by Burt Rutan.

Probably the best way to show the advantages of the composite structure is to compare it to a contemporary structure. The illustration (Fig. 5-11) compares the wings and canard of the VariEze homebuilt to a typical all-aluminum set of wings and horizontal tail. The comparison is equal in terms of utility, since both provide all lift, pitch control, roll control, and flap. Both are detachable for trailering. Both are intended for homebuilt construction and are of approximately the same size. Note that although the wings are approximately the same aspect ratio and weight, the composite wing's ultimate strength is 2.6 times that of the metal wing. The canard strength is also much greater, but that comparison is unfair since the canard has three times the aspect ratio of the horizontal tail. Using accepted calculation methods, the expected fatigue life of the composite wing is over nine times greater than the metal wing in their respective applications. The metal structure suffers a performance loss due to airfoil contour deforming and wrinkling under load.

The construction times are a conservative estimate based on interviewing several builders. This is not really a fair comparison since the VariEze was specifically designed for ease of construction. However, an indication of why the metal structure takes so much time lies in the large number of parts. The metal flying surfaces shown have 258 structural parts, 182 pieces of hardware, and over 1500 rivets.

I've been asked if the non-conductive composite structure needs special protection from lightning strikes. After researching this, I have concluded that there is no need for concern about structural failure due to lightning. Composites have been used in military aircraft for many years and recently they have incorporated precautions including conductive mesh laid up with the laminates because lab tests have shown that sweep lightning or direct lightning can break down the epoxy matrix causing delamination. This has not proven to be the case in actual experience with all-composite aircraft, however.

In checking with several very experienced sailplane pilots (one who holds the altitude record, who has flown sailplanes in every continent of the world, and who has had two direct lightning strikes himself) and with an experienced glass sailplane engineer and repairman (who has directed repair and modification of glass composite sailplanes for the last 12 years in Australia, Europe, and U. S. A.), I have found consistent and interesting reports. First, the competition sailplane spends a greater percentage of its life in thunderstorms than the average light plane. The competition sailplane trives on lift, and thunderstorms have the vertical gradients the sailplane pilot seeks. Recently cloud flying in competition has been outlawed in the U. S. and in Australia for one reason—accidents due to collisions of sailplanes with each other. Cloud flying and thunderstorm exposure is still prevalent in Europe. There are horror stories of severe thunderstorm flying—the most popular being the world championship soaring contest in Yugoslavia, in which as many as 60 sailplanes (most of them glass composite) were in a severe lightning/thunderstorm system at the same time. One, a wood ship, was destroyed by heavy hail and another driven into the ground by heavy rain and turbulence. Several wood ships sustained mild lightning damage but none of the glass composite aircraft were affected.

The sailplane repair engineer interviewed told of several cases of damage to wood sailplanes due to lightning strikes. These included minor, repairable wood structural damage, seized control system bearings due to the electrical discharge, holes in metal pushrods and severed rudder cables, none preventing a safe landing. But, more important, I could find no one who knew of any lightning strike damage on any glass composite aircraft. In one case a glass ship landed in high voltage electrical transmission lines. Extremely high current arcing occurred between the rudder pedals and metal structure behind the pilot. The flash was so bright that the pilot suffered eye injuries, yet no damage occurred to any glass structure.

Due to the extensive thunderstorm flying experience with no history of damage, the glass composite sailplane pilot routinely flies in lightning conditions without concern. This is particularly true in Europe where IFR competition is still legal. There are no special lightning protection devices installed in any of the glass composite sailplanes, including those intended for competition in IFR conditions.

The VariEze structure should be even less susceptible to

critical damage than the average glass sailplane—its highly redundant monocoque structure can sustain major damage and still support normal flight loads. Its rudder cable and position light wires are located such that if they are vaporized by a direct strike, the adjacent affected glass structure can be destroyed without destroying the structural integrity of the wing.

According to the NASA source interviewed, a wood aircraft is more susceptible to major lightning damage than a glass aircraft. This is due to the water moisture in the wood being immediately transferred to steam causing an explosion—the same reason a tree explodes when hit by lightning.

I didn't really mean to use this many words to explain that lightning is not a problem, but I felt I had to answer an uninformed report which inferred that flight near lightning meant certain total disintegration of the structure.

While we believe the skill required to build a VariEze is less than that required to weld or to form sheet metal, the method must be learned and we have the responsibility to educate the homebuilder. Thus, the plans will be much more than construction drawings—they will also include a complete, highly detailed, photo-illustrated construction manual and an education in composite methods. After the plans are finished, we will schedule seminars in major cities, in which all phases of construction will be demonstrated. Detailed construction articles will also be written for publication. Since the advantages of the structure are so evident, we believe it will soon be a "standard" and we want to be sure it gets off to a good start. Thus, we will be strongly against any modifications or materials substitutions until they are thoroughly tested and proven.

Originally we had planned to market the VariEze the same as the VariViggen: i.e., market only the plans, construction manuals, formed materials (canopy, cowling, wheel pants and landing gear) and machined parts. Several problems with this plan became apparent. Raw materials shopping can be quite a chore for the homebuilder, having to buy a few nuts and bolts here, epoxy there, etc. Substitutions of the wrong materials can result in extra work, extra weight or inadequate structural integrity. I feel that we have a responsibility to assure the complete success of the introduction of composites to the homebuilder by not only completely educating him on the methods of construction, but assuring him of being able to obtain the correct materials.

Further, we are a small organization, presently incapable of handling a large volume of materials and having no ambition to develop a large materials distribution organization. If we did, it would no doubt take years to reach the efficiency and productivity of the best companies now in this business. Therefore, we have contracted with present aircraft materials distributors and manufacturing firms to produce and distribute all the manufactured parts and all the raw materials in the VariEze bill of materials. The companies were selected based on their excellent record of delivery and customer satisfaction. They have all been in this business for a long time and will require very little time to begin volume delivery of kit materials. We are working directly with them, supplying engineering specifications of the materials and assisting with quality control. We will continue to work directly with them to supply and support any changes required in the manufactured items and raw materials.

Thus, we will directly market plans and construction manuals only, and will assist the homebuilder in the use of all materials purchased from our authorized distributors. The homebuilder will then be able to receive materials directly, from the best source, very soon after the inception of the program.

Volume orders of raw materials have already been placed. A major manufacturer is now weaving the special bidirectional and unidirectional fiberglass cloth to our specification. Final planning is underway to mass produce the manufactured items—they will be on the shelf in quantity before any customer spends his money for them. We currently plan to make the plans, construction manual and composite structures handbook available in May 1976. All materials and all components should be available at that time.

COMPOSITE CONSTRUCTION IN DETAIL

Most readers of this book are not going right out to buy the "makings" of a composite airframe. What we show here is a brief overview of how Rutan's designs are fabricated. Hopefully, this will give potential builders an inkling of what's in store for them should they eventually build their own fiberglass and foam airframe.

Much of the instant popularity of the VariEze and Rutan's follow-on designs comes from a structure made of glass/foam without molds. Homebuilders using this relatively new type of structure are encouraged to do their homework with a primary book entitled *Moldless Composite Sandwich Homebuilt Aircraft Construction*.

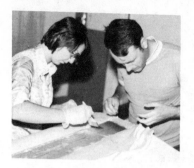

Fig. 5-12. Builders Chuck and Barb Banks wetting out the fiberglass. She's brushing epoxy and he's trimming the glass cloth (courtesy RAF).

Excerpts from this Rutan booklet give a basic concept of glass construction. Readers who like what they see and really want to get into a serious study of glass homebuilding should contact the Rutan Aircraft Factory for their copyrighted booklet. What follows is just a "tester" from the briefing manual and is reproduced with permission of the copyright owner. (Figures 5-12 through 5-22 show some of the steps of composite construction).

Glass

The most basic structural material in your VariEze is glass cloth. Glass cloth is available commercially in hundreds of different weights, weaves, strengths, and working properties. The use of glass in aircraft structures, particularly structural sandwich composites, is a recent development. Very few of the commercially available glass cloth types are compatible with aircraft requirements for high strength and light weight. Even fewer are suitable for the hand-layup techniques developed by RAF (Rutan Aircraft Factory) for the homebuilder. The glass cloth used in the VariEze has been specifically selected for the optimum combination of workability, strength, and weight.

Fig. 5-13. Hot-wire saw following a template to cut wing core (courtesy RAF).

163

Fig. 5-14. Happiness is a perfect layup for a perfect wing (courtesy RAF).

Fig. 5-15. Using brush and squeegee to wet out and smooth the shear-web layup (courtesy RAF).

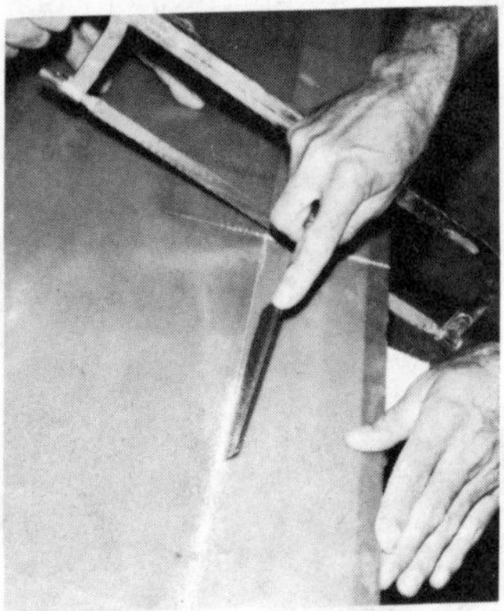

Fig. 5-16. Removing aileron from the wing (courtesy RAF).

Fig. 5-17. Knife trimming the edges before full cure (courtesy RAF).

Fig. 5-18. A formal wing static test with lead bags to apply a 10-G load (courtesy RAF).

The glass cloth in your VariEze carries primary loads, and its correct application is of vital importance. Even though doing your glass work correctly is important, this doesn't mean that it is difficult; in fact, it's VariEze!

Two types of glass cloth are used, a bi-directional cloth (BID) and a unidirectional cloth (UND) (Fig. 5-23). BID cloth has half of the fibers woven parallel to the selvage edge of the cloth and the other half at right angles to the selvage, giving the cloth the same strength in both directions. The selvage is the woven edge of a bolt of fabric. UND cloth has 95 percent of the glass volume woven parallel to the selvage, giving exceptional strength in that direction and very little at right angles to it (Fig. 5-24).

Fig. 5-19. An informal test with eight people applying a 7-G load (courtesy RAF).

165

Fig. 5-20. The Long-EZ wing core in its jig. It is being checked for twist before skinning (courtesy RAF).

BID is generally used as pieces which are cut at 45-degree angle to the selvage and laid into contours with very little effort. BID is often applied at 45-degree orientation to obtain a desired torsional or shear stiffness. UND is used in areas where the primary loads are in one direction and maximum efficiency is required, such as the wing skins and spar caps.

Multiple layers of glass cloth are laminated together to form the aircraft structure. Each layer of cloth is called a ply and this term will be used throughout the plans.

Newcomers are urged to cut a square play of BID and see how easy it is to change its shape by pulling and pushing on the edges. Cut a square with the fibers running at 45° and pull on the edges to shape the piece. It helps if you make fairly straight cuts, but don't worry if your cut is within 1/2 inch of your mark. As you cut BID it may change shape, just as the square ply that you are experimenting with does when you pull on one edge. Plies that distort when cut are easily put back into shape by pulling on an edge.

The fiber orientation called for in each materials list is important and shouldn't be ignored. UND is characterized by the major

Fig. 5-21. Landing gear attachment layup custom fits landing gear strut to fuselage (courtesy RAF).

Fig. 5-22. Basic Long-EZ fuselage after drying at Mojave (courtesy RAF).

fiber bundles running parallel to the selvage and being much larger than the small cross fibers which run at right angles to the selvage. In BID, the cross fibers are the same size as those running parallel to the selvage, giving BID an even "checkerboard" appearance. BID is commonly used for plies cut at 45° to the selvage. Your tailor would call this a "bias" cut.

Epoxy

In recent years, the term "epoxy" has become a household word. Unfortunately, "epoxy" is a general term for a vast number of specialized resin/hardener systems, the same as "aluminum" is

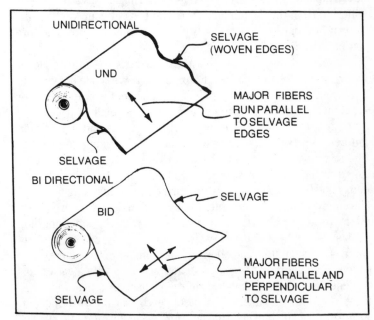

Fig. 5-23. Unidirectional and bidirectional fiberglass cloth.

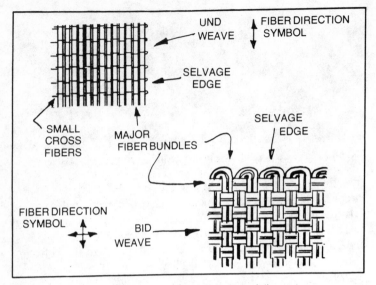

Fig. 5-24. Unidirectional and bidirectional fiberglass cloth.

a general term for a whole family of specialized metal alloys. Just as the "aluminum" in the spar of a high-performance aircraft is vastly different from the "aluminum" pots and pans in your kitchen, the "epoxy" in your VariEze is vastly different from the hardware store variety (Fig. 5-25).

Epoxy is the adhesive matrix that keeps the plies of load-carrying glass cloth together. Epoxy alone is weak and heavy. It is important to use it properly so that the full benefits of its adhesive capability are obtained without unnecessary weight.

An "epoxy system" is made up of a resin and a hardener tailored to produce a variety of physical and working properties. The mixing of resin with its hardener causes a chemical reaction called curing, which changes the two liquids into a solid. Different epoxy systems produce a wide variety of solids ranging from extremely hard to very flexible. Epoxy systems also vary greatly in their working properties; some are very thick, slow-pouring liquids, and others are like water. Some epoxy systems allow hours of working time and others harden almost as fast as they are mixed. A single type of resin is sometimes used with a variety of hardeners to obtain a number of different characteristics. In short, there is no universal epoxy system; each has its own specific purpose and while it may be the best for one application, it could be the worst possible in another use.

168

The RAF-type epoxy systems used in the construction of your VariEze (Fig. 5-26) are tailored for the best combination of workability and strength, as well as to protect the foam core from heat damage and solvent attack. These systems are also low in toxicity to minimize epoxy rash. These epoxies are not similar to the common types normally marketed for fiberglass laminating.

Three different systems are used in the VariEze for three different types of work: a slow-curing system, a fast-curing system, and a five-minute system. The very fast curing (five-minute) system is used much like clecos are used in sheet metal construction (or clamps in woodwork) for temporary positioning. Five-minute is also used in some areas where high strength is not required, but where a fast cure will aid assembly.

As an epoxy system cures, it generates heat and in some areas the heat buildup of a medium or fast curing epoxy system is unacceptable. Where this is a potential problem, a slow curing system is used. Slow cure epoxy is always used with styrofoam

Fig. 5-25. Bottles of epoxy and hardener lie beneath a homebuilt balance designed by Burt Rutan. A 3-ounce paper cup is used for hardener, at the left of the balance, while an 8 to 10-ounce cup handles the liquid resin. VariEze builder Marshall Gage of Whittier, California, studies detailed construction drawings before taking his next step.

Fig. 5-26. Marshall Gage, Whittier, California, is building his VariEze in his 18 × 20-foot garage. He can attach only one wing at a time in this small space. Gage stated that he chose the VariEze concept because he didn't need a machine shop to build it.

where heat can melt the foam away and ruin the joint. In other areas where heat buildup isn't a problem and a faster cure is desirable, a fast curing system is used. Both the fast and slow cure epoxies will cure to a firm structure at room temperature within one day. Complete cure takes 14 days.

Epoxy resin and hardener are mixed in small batches, usually six ounces or less, even in the larger layup. The reason for small batches is that, in large batches, as the hardening reaction progresses, heat is generated which speeds the reaction which causes even more heat, which ends up in a fast reaction called an exotherm. An exotherm will cause the cup of epoxy to get hot and begin to thicken rapidly. If this occurs, throw it away and mix a new batch. The small volume batch avoids the exotherm. For a large layup, you will mix many small batches rather than a few large ones (Fig. 5-27). With this method you can spend many hours on a large layup using epoxy that has a working life of only a few minutes. If the epoxy is spread thin as in a layup its curing heat will quickly dissipate and it will remain only a few degrees above room temperature. However, in a thick buildup or cup, the low surface area to mass ratio will cause the epoxy to retain its heat increasing its temperature.

Fig. 5-27. Here's just about everything a builder needs to complete a composite airplane. Mike Melvill, left, and Dick Rutan used this work bench to prepare the materials for two Long-EZs at Mojave. Note that the epoxy is mixed in many small batches, rather than a few large ones.

This results in a faster cure causing more heat. This unstable reaction is called an exotherm. Exotherm temperatures can easily exceed the maximum allowable for foam (200°F) and damage the foam-to-glass bond.

Microspheres

Microspheres are a very light filler or thickening material used in a mixture with epoxy. Micro, as the mixture is called, is used to fill voids and low areas, to glue foam blocks together, and as a bond between foams and glass skins. Several different types of microspheres and microballoons are available commercially. The quartz-type supplied by RAF distributors is lighter and cheaper than most common types. Microballoons must be kept dry. If moisture is present, it will make them lumpy. Bake them at 250°F, then sift with a flour sifter to remove lumps. Keep the microballoon container covered.

Micro is used in three consistencies: a *slurry* which is a one-to-one by volume mix of epoxy and microspheres, *wet micro* which is about two-to-four parts microspheres by volume to one part epoxy, and *dry micro* which is a mix of epoxy and enough microspheres to obtain a paste which will not sag or run (about five parts to one by volume). In all three, microspheres are added to completely mixed epoxy.

Flox

Flox is a mixture of cotton fiber (flocked cotton) and epoxy. The mixture is used in structural joints and in areas where a very hard durable buildup is required. Flox is mixed much the same as dry micro, but only about two parts flock to one part epoxy is required. Mix in just enough flock to make the mixture stand up. If "wet flox" is called out, mix it so it will sag or run.

When using flox to bond a metal part, be sure to sand the metal dull with 220-grit sandpaper and paint pure mixed epoxy (no flox) on the metal part.

Bondo

The term "Bondo" is used as a general term for automotive polyester body filler. Bondo is used for holding jig blocks in place and other temporary fastening jobs. We use it because it hardens in a very short time and can be chipped or sanded off without damaging the fiberglass. Bondo is usually a dull gray color until a colored hardener is mixed with it. The color of the mixture is used to judge how fast it will set. The more hardener you add, the brighter the color of the mixture gets and the faster it hardens. This simple guide works up to a point where so much hardener is added that the mixture never hardens. Follow the general directions on the Bondo can for fast setting Bondo. Mixing is done on a scrap piece of cardboard or plywood (or almost anything) using a hard squeegee or putty knife. A blob of Bondo is scooped out of the can and dropped on the mixing board. A small amount of hardener is squeezed out onto the blob and then you mix to an even color. You will mix the blob for about one minute. You will then have two to three minutes to apply it before it hardens.

Foam

Three different types of rigid, closed-cell foam are used in your EZE (and several densities). A low density (2 lb/ft^3), blue, large-cell styrofoam is used as the foam core of the wings, winglets and canard. The blue foam is exceptional for smooth hot wire cutting of airfoil shapes. The large cell type used provides better protection from delamination than the more commonly used insulation-grade styrofoams.

Low density (2 lb/ft^3 green or light tan) urethane foam is used extensively in the fuselage and fuel tanks. Urethane foam is fantastically easy to carve and contour and is completely fuel proof. The urethane used is U-Thane 210 or equivalent.

PVC foam in medium and high densities is used in fuselage bulkheads and other areas where higher compressive strength is required. The light red PVC is 6 lb/ft^3 and the dark red is 16 lb/ft^3. We considered using the "fire resistant" brown urethane instead of the green 2-lb urethane, but found its physical properties, fatigue life and fuel compatibility to be much lower than the urethane supplied to VariEze builders. Do not confuse styrofoam with white expanded polystyrene. Expanded polystyrene is a molded, white, low density, soft foam which has the appearance of many spheres pressed together. This is the type used in the average picnic cooler. It disappears immediately in the presence of most solvents, including fuel, and its compression strength and modulus is too low.

All three types of foams, PVC, urethane and polystyrenes are manufactured in a wide variety of flexibilities, densities and cell sizes. Getting the wrong material for your airplane can result in more work and/or degraded structural integrity.

Sun damages foam. Keep covered.

Hot Wire Cutting

The airfoil-shaped surfaces of your VariEze are formed by hot wire cutting the blue styrofoam of 1 lb ft^3 density. The hot wire process gives airfoils that are true to contour, tapered, properly twisted, and swept with a minimum effort and the simplest of tools.

The hot wire saw is a piece of stainless steel safety wire, stretched tight between two pieces of tubing. The wire gets hot when an electrical current passes through it and this thin, hot wire burns through the foam. The blue foam used in your flying surfaces was selected for a combination of reasons and its hot wire cutting ability was one of them. Use only the recommended materials.

Urethane Foam Shaping

One of the real treats in the construction of your VariEze will be shaping and contouring urethane foam. Urethane is a delightful material that shapes with ease using only simple tools. A butcher knife, old wire brush, sandpaper, and scraps of the foam itself are the basic urethane working tools. A vacuum cleaner is convenient to have handy since working urethane produces a large quantity of foam dust. The knife is used to rough cut the foam to size. The knife needs to be kept reasonably sharp; a sander or file is an adequate knife sharpener since it's a frequent task and a razor edge isn't necessary. Coarse grit sandpaper (36 grit) glued to a board is used for rough shaping outside contours.

Keep your shop swept reasonably well. The foam dust can contaminate your glass cloth and your lungs. Use a dust respirator mask while carving urethane. Try not to aggravate the better half by leaving a green foam dust trail into the house.

Glass Layup

The glass layup techniques used in your VariEze have been specifically developed to minimize the difficulty that glass workers have traditionally endured. Most of the layups that you will do will be on a flat horizontal surface without the molds, vacuum bags, and other special equipment that are common in glass work. The layups that you do will all cure at room temperature; no ovens or special heating is required. If you have suffered through a project that requires you to build more molds and tools than airplane components, then you are in for a real treat.

The techniques that you will use are VariEze, but they still need to be done correctly.

Before you get started with a layup, plan ahead. Some major layups take several hours and before getting your hands in the epoxy, it's a good idea to make a pit stop at the restroom. Do not start a large layup if you are tired. Get some rest and do it when fresh. It's best to have three people for any large layup—two laminators and one person to mix epoxy. Be sure the shop is clean before you start. Take the recommended health precautions using gloves or barrier skin cream. Get your grubby old clothes on or at least a shop apron.

If you use skin barrier cream, the epoxy and cream will wash off easily with soap and water. When you get epoxy on unprotected skin, Epo-cleanse is used to remove the epoxy. Both of these products are available through RAF distributors and are listed in the bill of materials. Once you are sure your skin is clean, wash again thoroughly with soap and water, even if your hands were protected with plastic gloves. If you get epoxy on tools or metal parts, clean them with acetone or MEK before the epoxy cures.

Quality Control Criteria

One of the unique features of the glass-foam-glass composite construction technique is your ability to visually inspect the structure from the outside. The transparency of the glass/epoxy material enables you to see all the way through the skins and even through the spar caps. Defects in the layup take four basic forms: resin lean areas, delaminations, wrinkles or bumps in the fibers, and damage due to sanding structure away in finishing. Resin lean areas are white in appearance due to incomplete wetting of the glass cloth with epoxy during the layup. Delaminations in a new layup may be due to small air bubbles trapped between plies during the layup. The areas look like air bubbles and are distinctly visible even deep in a cured layup.

The following is a listing of the "critical areas"—the portions of the VariEze that must meet all the inspection criteria:

☐ Center section spar—entire outside skin and spar caps.

☐ All portions of the fuselage within 10″ of the engine mounts and canard lift tab attachments.

☐ All control surfaces.

☐ All flying surfaces in the shaded areas shown plus all overlaps at L.E. & T. E.

Major wrinkles or bumps along more than 2″ of chord are cause for rejection in the wings, canard and winglets, particularly on the top (compression side). This does not mean you have to reject the whole wing—anything can be repaired by following the basic rule: remove the rejected or damaged area and fair back the area at a slope of 1″ per ply with a sanding block in all directions. By watching the grain you will be able to count the plies while sanding. Be sure the surface is completely dull and layup the same plies as you removed, plus one more ply of BID over the entire patch. This will restore full strength to the removed area. Use this method to repair any area damaged for any reason—inadvertent sanding through plies during finishing, taxiing a wing into a hangar, etc (Fig. 5-28).

Health Precautions

If you work with epoxy on your bare skin, you can develop an allergy to it. This "sensitization" to epoxy is an unpleasant experience and is to be avoided. You generally have to get epoxy on your unprotected skin to become sensitized. If you use a protective barrier skin cream like Ply No. 9 (available from VariEze distributors) or disposable plastic medical examination gloves (also available from VariEze distributors) the allergy can be avoided.

Fig. 5-28. Rutan's composite construction makes it relatively easy to patch areas with the resultant structure as strong as before the "patch." Note rework on the front of this VariEze fuselage as Marshall and Rachel Gage inspect a nose cover panel.

The barrier skin cream also allows you to clean up with soap and water after a layup.

The RAF epoxy systems are low toxicity (SPI=2). However, many people (about 5 to 7 per cent) are sensitive to epoxy to a great extent and thus will find it impossible to build their airplane without extensive skin rash, facial swelling, etc. These people can get some help by using doctor-prescribed anti-allergy medicines and/or by using elaborate masks/multi-gloves, etc., to reduce exposure; however, in many cases the allergy is sufficiently strong to preclude their ability to make layups. Remember to *always* use skin protection; never let epoxy come in contact with bare skin, even if you have no reaction to it. Sensitivity is accumulative, such that you may later develop an allergy unless you protect your skin.

Chapter 6
Engine, Engine,
Who's Got An Engine?

Reliable, affordable, efficient powerplants have been the goal of aircraft designers since day one. The history of aviation has many examples of fine, innovative designs that literally never got off the ground because a suitable powerplant was not available at the time the design was developed.

Rutan's development of the VariEze faced similar problems. While there are many VariEzes flying with VW engines (Fig. 6-1), a motorcycle engine and various cut-down automotive engines, Rutan had consistently urged his builders to stick with proven aircraft engines until a proven, dependable automotive or other design is available. Early in the program, Rutan singled out the popular 100-hp Continental 0-200 as being the best power-package for his design (Fig. 6-2). This reliable, high-production engine has powered the Cessna 150 and several other trainers so that used and rebuilt engines are more-or-less readily available. Older Continental engines ranging from the A65/A75 through the C85/C90 have been used successfully.

Through newsletters, lectures and flight demonstrations, Rutan has stressed engine reliability. The designer and his know-ledgeable builders shared valuable engine experience, just as they did in the building and in the flying of these homebuilts.

Concerning the pusher engines, Rutan said, "As you engine experts know, the Continental 0-200 (100 hp), C85 and C90 engines have a special crankshaft for an FAA-approved pusher installation. These special cranks are rare and expensive. We don't believe that these special parts are necessary for the VariEze. The

177

Fig. 6-1. VW installation in early VariEze.

difference between the "pusher" 0-200 B and the tractor 0-200 A is a reinforced flange to take the high static thrust loads that you find in amphibian type or the slow aircraft. The O-200, C85, C90, C75, A80, A75 and A65 crankshafts are almost identical (not inter-changeable) and the A65 engine is approved as a pusher without modification. Because of the fixed-pitch prop, designed for 200-mph cruise, the thrust loads on the 100-hp 0-200 A are lower than they are on the 65-hp A65 in a 'normal' installation."

After 150 hours of flight testing in N4EZ, Rutan reported that there had been no measurable increase in crankshaft end play which would reflect thrust bearing wear. "Everything looks good for the 0-200 A on the VariEze aircraft. Do not conclude from this that the "A" engine is suitable on other pusher aircraft."

At this time, A. C. Boyle, an EAA Chapter Designee and A&P rated mechanic, passed along the following suggestions for build-ers who might plan to overhaul or service small Continental engines.

"If you are overhauling an A65, convert it to A75. The conver-sion is covered by Continental service bulletin M47-16, revised 9/25/68.

"Install the 100 octane valve conversion when overhauling any of the small Continentals.

"Pistons, rings, valves, valve guides, rocker arm bushings, rocker arm pivot pins, and bosses on the cylinder heads should all meet *new tolerances,* not service limits, if you expect to get the full TBO.

"Never have crankshaft grinding done by an automotive shop. Stick with an FAA approved regrind shop. The bearing journals on aircraft crankshafts have specific requirements that auto machine shops generally don't have the equipment to meet.

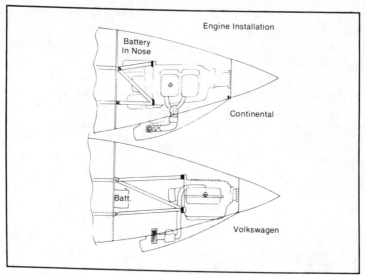

Fig. 6-2. Sketch of rear-mounted engine installation for both the Continental and VW powerplants (courtesy Rutan Aircraft Factory).

"Have your inspection (Magnaflux & Zyglo) and machine work (grinding and plating) done by *competent* aircraft machine shops. Don't take chances with amateurs.

"If you plan to store your newly overhauled engine over 90 days, don't test run it until you're ready to put it in service. The combustion products from the run-in will cause corrosion internally. If long-term storage is planned, oil all internal surfaces on assembly, plug all openings (breather, oil pressure port, intake, and exhaust openings), and install desicant spark plugs.

"Follow the service manual's recommended run-in procedure.

"Regarding the use of 100LL fuel in the low compression small Continentals. If adequate (100 octane) valves are installed, plugs cleaned every 100 hours, and oil changed every 25 to 30 hours, no appreciable degradation of engine dependability should be experienced as a result of 100LL use."

As the VariEze project developed. Rutan developed a perspective that he passed along to his builders.

"We are probably going to hear a lot of anguished cries from would-be engine developers, but we are taking a hard line on 'other' engines in the VariEze. We have had many calls and letters from people wanting to install all kinds of converted boat, snowmobile, auto, turbocharged trash compactor, etc., engines in a VariEze.

Also, there is an interest in all kinds of unproven modifications to the VW (fuel injection, turbocharged, electronic ignition, etc.).

"Frankly, we're scared stiff by this. Aircraft engine development is a very risky, horribly frustrating and enormously expensive business. Please don't kid yourself and think your new engine conversion isn't going to fail a few times during initial flight testing. Even a professionally trained, educated, and experienced engineering organization with a barrel of money can't do these things, so don't try it in your garage. Moreover, don't believe anybody who says he can do it for you, unless he can show you excellent maintenance records taken during hundreds of hours of *flying* with the engine.

"We are very much afraid that if a lot of homebuilders start trying to develop new engines on homebuilt airplanes, that EAA's accident record will look horrible. Doing engine development on an amateur-built airplane hurts every one of us by further endangering the lenient rules that we now have. Please don't do it.

"This isn't to say that some very good engines aren't hiding out there, waiting to be developed for aircraft use. We wish the best of luck to those who have the funding, ability and ambition to do the job well. Doing an engine development job well implies that you have the professional ethics not to endanger the hard-won privileges of others."

EXPERIMENT WITH A CUB—IF YOU MUST

"Now, if you have an engine that looks good to you and you really want to prove it out for aircraft use, here's what you do: fly it!

Fig. 6-3. Experiment with a Cub if you must. Rutan recommends strongly that engine development be left to the experts. The Cub has significantly lower landing speeds than the canards, reducing risk in case of a failure.

Fig. 6-4. VW hanging on half-finished VariEze prototype.

There is no substitute for flight experience. Not in a *homebuilt*, though! Get yourself a Cub or Champ that is a very forgiving airplane, easy to land safely in a pasture (Fig. 6-3). You are going to make several emergency landings, so plan on it. If things really get bad and you have to plant your test vehicle in the trees, then for the FAA, it's just another Cub that crashes—not a *homebuilt*. Also, you can buy another Cub and get your test program rolling again, quickly. If you had used a homebuilt, you would have to build another airplane instead of getting on with your engine development work.

"Once you get your new engine working like you think it should, fly the pants off it; maintain *detailed* maintenance records, and find out just how well it really holds up over a *full overhaul period*. Find out how much it really costs you to fly each hour, considering initial cost, operating costs, maintenance, replacement, and everything else. We once participated in a 'low cost' engine program where the initial engine cost was less than 10 percent of the 100-hp Continental, but taken hour-for-hour of service, the cheap engine costs more than eight times as much! Remember, an aircraft engine is the very definition of dependability and reliability. An aircraft engine must tolerate abuse and still keep pumping along.

"Right now there is a promising looking engine powering a Cub that is being considered for the VariEze. The developer was originally going to do his development testing in a homebuilt Cassutt racer. Fortunately we were able to talk him into using the Cub. During the initial flights, at least four precautionary landings had to be made. This is perfectly normal in initial flight testing of

181

new engines. In the Cub, it was no sweat! In the Cassutt (high wing loading, fast landing, high rate of descent), it might have meant a broken airplane, possible injury to the pilot, and another statistic to hurt our EAA safety record. These fellows are to be congratulated on a very sensible, professional and ethical test approach."

A comparison between the Continental 0-200 and the original VW in N7EZ (Fig. 6-4) was detailed. Rutan compared the cooling of the two engines as follows:

"Cooling on the 0-200 has been excellent. Ground cooling is better than on most factory-builts. At a recent fly-in, we had to sit in a long line on a hot ramp awaiting takeoff for over 30 minutes. Many of the factory-builts had to shut down to avoid overheating, but N4EZs temperatures stayed under the normal values for cruise.

Fig. 6-5. Ken Swain, Travis AFB, California, has a scimitar prop on his Continental-powered VariEze. He turns this combination at 2750 rpm and wants 3000 at sea level.

"The 0-200 Continental engine (Fig. 6-5) has been trouble-free, requiring no modifications, adjustments or unusual maintenance. This has not been true of our VW installation on N7EZ. We have flown the VW VariEze a total of about 280 hours (two different engines), which is a lot of flying for one year for a VW homebuilt. We have had its cowling off an average of once per five flights, making minor carburetor adjustments, trying to find oil leaks, adjusting or repairing valves, cleaning plugs, checking the magneto coupling, tightening loose bolts, etc., etc. The VW-powered Vari-Eze has never had an inflight power failure, but it has twice had to be landed within a few minutes or it would have had a failure. Once it was due to low oil pressure; another time it was due to an impending failure of the prop hub/extension. We have conducted an informal survey and found that our VW experience is quite similar to others who have high time on VW aircraft conversions. For this reason and because of the high cost of operating these engines, we are not preparing the section on VW engine installations at the present time. This section should be made available later as we gain some more reliability experience with the engines and after the results of some further tests.

"Those of you who plan to use a VW do not have to delay your project. The engine installation is done during Chapter 23 of Chapter I. If you can't find an engine or haven't decided which type you plan to use, we suggest you get *only* Section I and go ahead and start building. Chances are, you will find the right engine at the right price before you get to Chapter 23.

"Frankly, when we went shopping for an 0-200, we couldn't find one. We let several people know we were looking. Then all of a sudden several weeks later, we had our choice of four 0-200's! When searching through *Trade-A-Plane* for engines, don't look only at the engine section. Quite often you can buy a wind-damaged Champ or Cessna 150 with a lot of time left on the engine for less than a used engine!

"We, of course, are not in the business of engine distribution; however, to give you an idea of the current market, we have made up a table (see Appendix), based on available prices from vendors, from *Trade-A-Plane*, and magazine ads. Cost per hour is basic engine cost, not including fuel and oil."

SAVE MONEY ON YOUR PURCHASE

"New engine cost is approaching the ridiculous. But, if you can find a partially run-out engine, it can be more of a bargain. The

Long-EZ has a Lycoming 0-235 that had 1400 hours, and it cost us $1500. It was installed without modification or repair. The Lycoming has a 2000 hour TBO, so we can expect 600 hours flying on this engine (3-4 years). When it is 'run-out'—2000 hours—it will still be worth what it cost us. Thus, it's truly a *low* operating cost engine (zero $)! This newsletter lists a 400 hour C90 at $2800. If you run it out (1200 hours left), then sell it for $1500, you have spent $1.08 per hour for engine. If you buy a new or newly overhauled engine at the high going cost and sell it for run out costs, you will spend $2 to $3 per engine hour. Remember the majority of you will take years to put a few hundred hours on an engine. So consider a ½ or ¾ runout as a bargain. Run an ad in *Trade-a-Plane*—it works!"

Not all engine projects end in success. This cryptic note was included in the *Canard Pusher*.

"Franklin engine 2A-120-C, 2 cylinder 60 hp, 30 hours SMOH. Rebuilt bendix mags and carb with pusher prop. $1900 FOB Mojave. This is the engine out of N7EZ original prototype. The aircraft engine is EAA museum bound and engine is available. This engine is NOT recommended for a VariEze. Ask for Dick (Fig. 6-6, 6-7)."

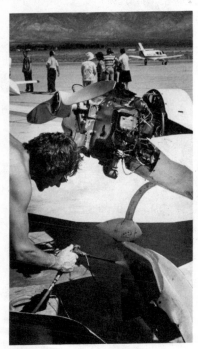

Fig. 6-6. Two-cylinder, 60-hp Franklin engine, a 2A-120-C, was installed in the prototype VariEze. Here Dick Rutan cleans the cowling on this installation that did not work well. "This engine is not recommended for the VariEze," said Dick.

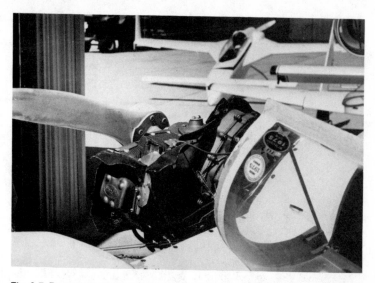

Fig. 6-7. Prototype VariEze N7EZ is scheduled for the EAA Museum, but will not be ferried back with this 2-cylinder Franklin powerplant. Vibration was excessive.

One of the more novel engine experiments concerned an early VariEze that Burt test flew at Mojave. The account of this first flight was supplied by the builder.

"We have talked to a number of 'high-time' VW flyers over the last few months and the story we hear is generally one of woe and poverty. The message that we get, even from successful bug operators, isn't very enthusiastic. One high-timer (800 hours) said that there were probably ten other guys who had suffered failures with VW's for each successful operator. He is using a Continental C-85 in his VariEze. Another bug operator we talked to said that in the 240 hours he was able to operate before failure, his VW cost him more dollars than a brand new 0-200 would have (at that time $3800), and he spent an hour with the cowling off for every hour he spent flying. The thing that scares us is that both of these gents are very competent engine mechanics with plenty of troubleshooting experience and a trained ear for problems before they get serious.

"On the other hand, we talked to a group of small-bore VW operators (36-50 hp) who had excellent service in motor gliders with very little hassle. Unfortunately, the 36 to 50-hp engines are too small to power a VariEze. Our present conclusion has to be that we can't encourage the use of VW conversions in the VariEze yet. We say *yet* because a number of development programs are underway and eventually the big-bore bug may become an aircraft powerplant, but don't hold your breath waiting. If we were to recommend a VW installation currently, we would expect to see a lot of unhappy builders, a high percentage of maintenance troubles, excessive costs, and possibly some accidents. Beyond basic engine durability the VW converters are faced with devising and supporting a service bulletin and airworthiness directive (AD) system to support the operator. Frankly, Continental and Lycoming would not have the success record that they do without the customer support effort in service and overhaul manuals, AD's and service bulletins.

"We've received a lot of flak over our selection of the Continental engines (Fig. 6-8) for the VariEze since most models have been out of production for years. The most common question is 'How about the Lycoming engines?' And this is our answer: 'They are too heavy. The 0-290, 0-320, 0-340 and 0-360 engines are *totally out of the question.'*

"The 0-235 models could be used only with some *strict* limitations. The normally equipped 0-235 is 242 pounds which is much too heavy both structurally and from C.G. considerations. If the 0-235 is *stripped* (mags and carburetor only remaining), its weight can be reduced to 211 pounds which is marginal but can be lived with (as is the Continental 0-200 with alternator but no starter); the 0-235 has some advantages in lower cost, and it is available in a 100 octane burning version."

Weight is critical in a small airplane like the VariEze. Rutan had some specific comments for builders who insist on installing an electric starter.

"Look at what you're doing to your airplane. First, you add a 16-lb starter to your engine; then you add a 25-lb battery in the

Fig. 6-8. Gary Hertzler of Tempe, Arizona, tops off his Continental-powered VariEze at Bullhead, Arizona. Note exhaust stacks mounted out the trailing edge of the wing. Reliability of the aircraft engine is far higher than automotive conversions.

nose to balance and power it; then you add six pounds of cable to connect the two both ways. (You can't ground to glass and foam.) Presto, you've added 40 pounds of empty weight that does nothing except in the first five seconds of a flight. A small seven-pound battery gives you everything you need for avionics and lights. For the privilege of pushing a button once each flight, you have reduced your useful load-carrying ability by 10%. Look at it this way: your starter-equipped airplane will go 330 miles *less* with the same takeoff weight as my hand-propped model.

"When N4EZ had 300 flight hours, its 0-200A had 1500 hours since major overhaul. It required no maintenance besides oil changes since it was installed in N4EZ two years ago. It rarely requires more than two flips to start. Oil temperature runs at 170°F. Cylinder heads run 420° in a long climb, 360° at cruise, and do not exceed 300° on a continuous hot-day ground run. It has never had a starter while on N4EZ and its alternator was removed a year ago. These two holes were covered with 3/16" aluminum plates. The logbook shows that the starter failed twice on the previous owner and the alternator once, costing him over $400! As you know, starters are not recommended on VariEzes. Only one has flown with a starter and that builder has since removed it. About ½ of the Ezes to initially fly with alternators have removed them also, to reduce weight. We strongly recommend that you first fly without starter and alternator. Add them later if you desire, but do initial flying as light as possible. Do use a carb accelerator pump or primer for easy starting.

"The following is a quote from Warren Curd, Raytown, Missouri: 'I originally had full electrical. The airplane flew well enough, but I had to carry weight in the nose; then when I started adding weight in the passenger seat I was less than satisfied. I finally took your advice and removed the starter, alternator, two solenoids, 22 feet of 1/0 cable, and heavy battery—weight reduction, 65 lbs! The airplane now flies and handles so much better I hardly can believe it. Takeoff and landing speeds and distance are greatly reduced; climb improvement is amazing. Take it from me—save yourself a lot of work later. Don't install a starter or generator in your Eze. Starting is easy and the trickle charger keeps the small battery sufficiently charged for radio needs.'

"Caution: Do not ever hand prop a VariEze (or any airplane) that does not have at least one functioning impulse mag. An impulse mag allows the plugs to fire at or slightly after top dead center; without an impulse mag it will fire up to 25° before top dead center, which can lead to broken thumbs at the very least. If you only have one impulse mag, be sure you select only that one until the engine is running."

PROPELLERS TO MATCH

Since engines and propellers must work well together, Rutan also monitored this development carefully.

"We have tested several propeller types and studied several others. Fortunately, the best prop has been the lightest and lowest cost—a fixed-pitch, all wood, two-blade with plastic leading edge for rain erosion protection. These are available through several vendors. The owner's manual will specify prop sizes, specifications and recommended vendors for all recommended engines.

"The three-bladed prop tested resulted in less take-off, climb and cruise performance as compared to the two-bladed props.

"Those of you who are in a rush to be flying a VariEze as soon as possible may not want to wait for an order for a prop, since props are one of the hardest things to get without waiting several months. Thus, we are providing sizes and manufacturers for the 0-200 props we've tested.

We haven't gotten too many requests for variable pitch/constant speed/adjustable props from our builders which is a tribute to their good sense and intelligence. However, for those few who have asked about them, this is why we are down on them: safety, cost, weight and maintenance. First, it is a very definite risk to install a variable pitch prop on a pusher aircraft. The development of a variable pusher prop for the VariEze could easily run hundreds of thousands of dollars and still be a failure. The cost of a proven variable prop, even if one were available, would be over a thousand dollars each. The lightest controllable prop would weigh about 25 pounds which would create a CG problem requiring ballast, further increasing the weight growth. The maintenance and upkeep required on a variable prop is unbelievable. Look through the FAA airworthiness directives for propellers, and you'll see what we mean. Even if you have money to burn, a full-time mechanic on salary, and are a hairy-chested test pilot type, you won't gain anything with your fancy prop. The added weight will limit your useful load. Even for gross-weight operation, the VariEze requires a larger airport for landing than for takeoff. Climb is excellent even with a fixed-pitch prop. Thus a variable pitch prop would not increase utility.

"Aircraft Spruce has a spinner specifically designed for the

VariEze. It has been in test on N4EZ for three months. Previous spinners have suffered cracks after a few hours. This one has two very stiff flanges that accurately fit the cone. It has a long pleasing shape, is 10.3" in diameter, and comes complete with forward airflow guide for best engine cooling and low drag. The flange is custom trimmed. I had given up on spinners until this one worked out so well."

One of the freedoms in homebuilt aircraft is that of experimentation and change (Fig. 6-9). However, most of Rutan's builders will follow the lead of the designer and stick to proven aircraft engines—at least 'til something better comes along.

THE FLYING HONDA

One of the more memorable efforts to obtain reliable power from an existing engine was developed by Jim Cowley and a team of engine experts from Santa Paula, California. With a glib, tongue-in-cheek approach, Cowley details the development of this water-cooled engine and airplane combination. The project began in 1976 and continued on, and on, and on.

"Once upon a time, we decided to fly with an automobile engine. At this time, the reason we wanted to undertake this endeavor escapes me, but it must have been a good one. However, we made one major mistake—we forgot to use our standard multiplication factor when working with aircraft. From past experience, we have found that when working on *any* aircraft, to determine the time and money needed simply decide how long the project will take. Next decide how long the project will take when everything goes wrong that can possibly go wrong. Add these two figures together and multiply by five. Ninety percent of the time you will be a little low, but still in the ballpark or airpark as in this case. The same basic formula works with money needed to do something on an aircraft. There is an exception to this, as with all mathematical formulae. When developing a flying automobile engine, the multiplication factor of five should have some zeros behind it. To date we don't know just how many zeros, but it could be several.

"At the start of this great project, around May '76, we had the use of a Cassutt airframe and thought it would make a good ground test stand. We fixed up a big slab of steel to use as a firewall, hung our first Honda reduction unit, and fired it up. It ran a short while. From this first attempt we learned several no-no's. One thing was how to get two crankshafts out of one. Everything was redesigned, run a bit more on the Cassutt test stand, and we began talking about

Fig. 6-9. Ken Swain, Nut Tree, California, hand props his Continental powered VariEze. Swain carved this scimitar prop himself. "If you can build the airplane, you can certainly build the prop," he commented. He is a USAF Captain flying C-141s from Travis AFB.

hanging the entire thing on something that would fly.

"Next stop, a trip to Mojave for a long talk with Burt Rutan about this worthy project. After Burt calmed down, he decided we were serious and that he might as well give us some practical advice and keep us alive as well. We ignored his first suggestion, but his next suggestion seemed like a good one. That was to hang the engine on a slow stable aircraft, and one whose flying characteristics we knew. This was real good because we had a J3 Cub available. It wasn't long before the J3 was stripped from the firewall forward. The new engine mount was built, accessories rearranged again to fit in the proper places under the new cowling, extra instruments put in, and we were off and flying. Everything was going well so we decided to head for Oshkosh.

"The Cub was given a nice bath, a radio installed, road maps obtained and all was ready for the big hop to Oshkosh for the '76 fly-in. Jim Kern went aloft, defying gravity and superstition, IFR (I follow route 66) with Jerry Kibler as ground support. A big red square was taped on top of the car so Jim could spot it easier from the air. After many stops, scheduled and unscheduled for gas and/or water, all arrived at Oshkosh in good shape, except for the car. It had to have the gear of the day set from underneath as Jerry was not able to shift gears from inside. This also made a problem of getting the thing in reverse and took a bit of pre-planning on parking. All went well at Oshkosh and the Honda Cub flew just about every day in the fly-by's. The trip back was a little longer due to hot headwinds and time out for side trips to visit friends, plus the normal scheduled and unscheduled stops for gas and/or water.

"The trip from Santa Paula, California, to Oshkosh and back to Santa Paula in a Honda Cub only takes 90.2 hours. We did learn a few things from that trip. From starting the engine to shutting it down, the Honda averaged 3.7 gal/hour. The entire trip was also made with a full rich mixture as we had no lean control on that carburetor. The Honda Cub had a 2.25 belt-driven reduction unit on it and we were using a shortened Bucker prop. The Uniroyal drive belt was 50 MM wide and lasted fine the entire trip. However, we had so many remarks at Oshkosh about that skinny little belt we have put a 80 MM wide belt on the new reduction unit. Most of the trip we used about 30 inches of manifold pressure for cruise. Takeoff was normally 40 to 45 inches. The highest we had the Cub on the trip was 12,000 feet and still had 30 inches at that altitude. We used a small Ray-Jay turbo charger on the Honda. The engine seemed to develop about 85 to 90 hp on the Cub, and we have since confirmed that on a dynamometer. At 5,500 rpm and 40 inches manifold pressure, it will produce about 90 hp. The Honda car will run down the road at 60 mph at 5,000 rpm, so this seems like a good figure to try for on the aircraft. At that rpm the engine is not running too fast or lugging down.

"Our next plan—install the Honda on the VariEze. This, of course, was quite simple. All we had to do was build a VariEze, modify the reduction unit to a 1.8 ratio, redesign the exhaust and intake manifolds and numerous other little odds and ends. This wonderous feat was accomplished in exactly four months. It was all a very calm operation, similar to the night we painted the aircraft. This was done at Santa Paula at midnight, January 13th, in the middle of the taxiway with the lights from three cars so we could see what we were painting besides ourselves. It all took about an hour to do the job, and it's really a lovely paint job—*if* you look at it through a camera lens at a distance of fifty feet or more. Hence, this is what we call our fifty-foot paint job. The wings and canard were taken to Mojave to be painted by our shop out there early on January 13th. [Cowley, Inc., makes vacuum formed plastic canopies at the Mojave Airport.] About 5 am on Janaury 14th, we loaded the fuselage on a pickup and headed for Mojave (97 miles away). The FAA inspector was scheduled to meet us at Mojave at noon. Shortly after noon on January 14, 1977, we had one VariEze Honda-Powered Aircraft, N344SP, inspected and ready to fly. Whoopee!

"The next several days were spent driving it all over the Mojave Airport, adjusting brakes, taxi tests, adjusting brakes, adjusting trim, adjusting brakes, etc. Finally discovered we had one small problem. The right disc was thick in spots and thin in spots. This varied as much as .040. This caused the brake to grab and that made the gear shimmy, and that broke the strap of the gear leg. Simple fix—grind the disc flat. Of course, we also had to cut a big hole in the bottom to remove the gear and put a new strap on the gear leg and then patch all that. After a while we were back to flying status and on with the taxi tests, and adjusting brakes.

"It was mutually decided by everyone (except Burt) that Burt should have the honor of flying the first plans-built, Honda-powered VariEze. After chasing him down, we strapped him tightly in the cockpit, wound it up and pointed him in the direction of the runway. We all stood back and with encouraging words and cameras pointed, watched this great flight. It actually took off, made turns at the proper times, and flew several times around the airport and came in and landed just like it was designed to do.

"Amazing! When we let Burt out of the cockpit, he was mumbling something about it not being right for us to have a quieter and smoother-running engine than he had. Jim Kern had the honor of next going around the patch with it. We then decided it had been a long day (January 19th) and a longer four months. We gave N344SP a friendly pat on the nose and put her to bed. Then we all went out to toast Burt, Hondas, VariEzes, Cowley, Inc., Kern and anyone else we could think of.

"Burt had been developing the spoilers and ailerons for the VariEze and thought we should add them at this time. We had to make some more modifications to the engine compartments to get all the tubes and fittings in the right places and just got back in the air when the "Mojave Breeze" started. We sat around for four months with the wind blowing from 30 to 100 knots hoping to get some performance figures on the flying automobile engine before taking it back to Santa Paula for the aileron modification. We finally decided we might as well bring the Eze back to Santa Paula, put the ailerons on, and hope the wind would let up a bit by the time we got it back to Mojave. The ailerons also mean pulling the engine and making some more modifications there. N344SP should also have a nice paint job inside and out at the same time. The engine and reduction unit should be changed to what we believe will be the final configuration. And as long as we are going to do that we might as well . . . Like any aircraft, when you put air in the tailwheel you might as well do this and that until eventually you have overhauled the engined or put on new fabric or some other dumb thing.

"How does the VariEze fly with a Honda Civic engine? We wish we knew. About all we can tell you is, it's very, very quiet and smooth. We have flown with an old T'Craft prop. It is in no way near the pitch and diameter we need for the best performance. This prop *was* 72" long, but on one of the trial runs the ship was over rotated and the prop was trimmed about an inch and the winglets about ½". We have also flown with a 66-66 prop made by Ted, but have been unable to get the rpm's we want with that. We do believe by changing the ratio of the reduction unit slightly and cutting this prop a bit shorter, we will be able to get fairly close to the right rpm and manifold pressure.

"To date we have around 250 hours on test stand, Cub and VariEze combined. This includes a long cross-country in the Cub. Our biggest problem has been cooling, both on the Cub and the VariEze. There is also the problem of time and money, and it has taken a lot more of both than we anticipated. We believe we will be able to keep the engine package just under 200 lbs with full electrical system, alternator, distributor, turbo, intake manifold, exhaust manifold, reduction unit, engine mount, and so on.

After the Honda Eze went to Santa Paula for ailerons, Beverly Cowley reported in February, 1980, that everyone who was interested in the Honda project was involved in other things. Since then, the poor Eze has sat as we seldom have time to work on it because of the canopy business. However, it has had the ailerons installed, new paint job inside and out and a Continental 90 installed. It needs only a new cowling installed and will be ready to fly again.

Chapter 7
Long-EZ—The Record Breaker

The Long-EZ (Fig. 7-1) design is the most popular homebuilt airplane yet to be developed. Within the first two months that the plans were available, 500 sets were sold! Sales continue at a record-setting pace and Long-EZs are sprouting up all over the world to continue the success record of the VariEze.

Rutan's initial announcement of the Long-EZ explained:

"We decided to design a new aircraft around the Lycoming 0-235 with starter and alternator. It would have unusually long range, thus the name 'Long-EZ.' It would have good forward visibility on landing and a lower approach and landing speed than the VariEze, making it more suitable for the low-proficiency pilot. The original configuration of the Long-EZ used VariEze wings placed on a center section spar that was four feet longer than a VariEze. The wings were swept more than a VariEze to support the heavier engine. It had a 'rhino' rudder on the nose and no control surfaces on the winglets. That aircraft, N79RA (Fig. 7-2), was built in four months in the spring of 79 and made its first flight in June, 1979."

The Long-EZ really started out in life as a training project for Mike Melvill (by then working for Burt Rutan) who completed the first VariViggen. As the project got underway, Burt decided to modify the VariEze design to accept the popular Lycoming 0-235 115-hp engine. Thus the fuselage was widened 2″ and beefed up to handle the extra weight; wingspan was upped to 26′ and the Long-EZ was soon in the air. Following first flights, extensive modifications were made in the design because it did not fly well.

Fig. 7-1. The Long-EZ comes in for a landing at Mojave with Dick Rutan giving instruction from the back seat. The original "rhino" rudder has been eliminated and the dive flap has not yet been installed.

As Burt explained, "Directional stability was weak; dihedral effect was excessive; adverse yaw was high; roll rate was sluggish; early airflow separation on the wing caused pitch instability at low speeds, and the stall speed was too high.

"During the next five weeks, N79RA made 51 flights, testing the effects of over 30 different modifications. Modifications included many configurations of wing leading-edge cuffs, wing fences and vortex generators. The winglet 'cant' angle was changed. The ailerons were rigged to various neutral positions. Some of the changes resulted in improvements in pitch stability and lateral-directional flying qualities. However, we were unable to improve the stall speed, landing attitude and roll rate to a satisfactory level. By August, we were convinced that to get the Long-EZ we really wanted, we would have to build an entire new aft wing.

"The new aft wing, first flown in October '79, had the following improvements:

☐ Less sweep.
☐ More area.
☐ A new Eppler airfoil similar to that on the Defiant.
☐ Longer ailerons.
☐ Improved winglet juncture to eliminate airflow separation at wingtip.
☐ Overlap-type wing attachment to centersection spar, allowing incidence adjustment.

"The new wing performed excellently in test. Approach and stall speeds were lowered. The lower approach and landing attitudes allowed 'full stall' landings with good runway visibility. Roll rate was superb. Directional stability was still weak and rudder power at the new lower landing speeds was inadequate. We then built new, larger winglets with rudders and removed the 'rhino' rudder from the nose. That configuration completed flight tests in December '79 with very excellent results in every respect.

"It has been shown to be resistant to departure during every conceivable stall entry, including tailslides. Its stability is firm even at maximum aft cg (obtained with a 120-pound pilot with starter, alternator and vacuum pump installed on the 0-235 Lycoming). Even though it has a wing area 41 percent greater than the VariEze and a 26 percent greater gross weight, it cruises at 183 mph at 75 percent—only 14 mph slower than the VariEze.

"The Long-EZ is now the recommended airplane for the 0-200 Continental and 0-235 Lycoming engines. Complete electrical systems, including starter and night lighting are approved."

The first sets of completed plans were taken by Dick Rutan and Mike Melvill to build twin Long-EZs (Fig. 7-3). They purchased the material package from Aircraft Spruce, rented a small section of a barracks building and spent every night for several weeks in the fabrication of two identical Long-EZs (Fig. 7-4).

"We're using this project to prove out the plans," explained Melvill. "By the time these two airplanes are complete, we should have every problem eliminated in our working plans. And besides that, we'll each have our very own aircraft."

LET'S BREAK A RECORD!

At 7:27 on the morning of December 15, 1979, Dick Rutan lifted the Long-EZ prototype off Mojave, California's Runway 12

Fig. 7-2. Dick Rutan with passenger Bob Weston in flight near Mojave. This represents the completed configuration of the Long-EZ with the new wing installed.

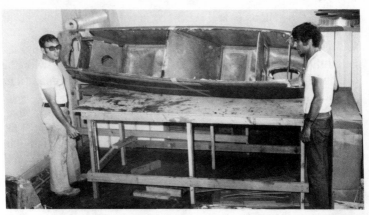

Fig. 7-3. Mike Melvill and Dick Rutan used the first set of plans to begin construction of two personal Long-EZ aircraft for their own personal use. The fuselages, shown here, took two weeks of after-hour and weekend work.

. . . and 33 hours, 33 minutes and 41 seconds later landed on 24 after covering an incredible 4800.28 statute miles, a new world's closed-course distance record for aircraft weighing between 1102 and 2204 pounds (Class C-lb).

[This narrative of the Long-EZ's record flight was prepared by Jack Cox with background information from the Rutan family, Burt, Dick, George and Irene. Excerpts are from Cox's article carried in the February 1980 *Sport Aviation* from which we continue to quote with thanks.]

The flight eclipsed the old record of 2955.39 miles held for the past 20 years by Jiri Kunc of Czechoslovakia. Rarely is a world's record exceeded by such a wide margin as Dick's 1844.89 statute miles.

The facts and figures of the flight are as follows:

☐ The aircraft—Rutan Long-EZ, N79RA, powered by a 108-hp Lycoming 0.235.

☐ Fuel on board at takeoff—143.6 gallons.

☐ Weight at takeoff—1904 pounds, 604 pounds over the normal gross of 1300 pounds. The take-off weight, however, was 300 pounds under the weight limit of the F.A.I. class—which means an additional 50 gallons of fuel could have been carried.

☐ The course—Mojave to Bishop, California and return equalled one lap. Fifteen laps were flown.

☐ The average speed for the flight was 145.7 mph.

☐ Average fuel flow was 4.17 gallons per hour, or about 35 miles per gallon.

197

☐ 3.75 gallons of fuel remained in the aircraft's tanks at the conclusion of the flight. 2.3 quarts of oil were consumed.

On December 7, Dick flew the dive/flutter tests and investigated the aft C.G. limits. Finding everything to agree with the computer, the design was frozen that day and, at last, preparations for the record attempt could begin.

Dick had already been working on an auxiliary fuel tank—one that would fit the rear seat area like a hand in a glove (Fig. 7-5). This was finished—in foam and glass, of course—and installed in the airplane. In short order other mods followed.

The fuel system was altered to route the flow from the tanks to the front cockpit and back to the engine. This was done because a Sears fuel flow meter was installed in the instrument panel to keep the extremely accurate tabs on gasoline consumption that would be absolutely essential for the successful completion of the flight.

Since about 14 hours of the projected flight would be in darkness, a lighting system had to be installed. Jack Gretta of Whelan made up a custom rig that could operate on just 6 amps. Included was a retractable landing light in the belly of the fuselage that would serve double duty, as you will soon learn.

An AC dyno from a Kubota tractor was installed. Including a small belt to the engine and the regulator, it only weighs about five pounds and produces 6 amps. (It has worked so well that RAF may offer installation drawings for homebuilders.)

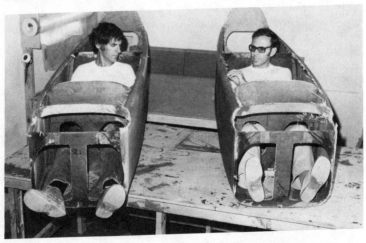

Fig. 7-4. The fiberglass fuselages for two new Long-EZs are checked for size by Dick Rutan, left, and Mike Melvill. Note rolled glass in the background that was used for construction.

The 604 pound over normal gross take-off configuration was too much for the fiberglass landing gear legs—they could handle the weight, but would spread out too much. Burt's fix was to install a sturdy cable bridle between the main gear axles to restrict their tendency to spread apart. There was no living with the drag this would produce, so a 1/16" cable was rigged up so it ran from a T-handle inside the cockpit, down through the open landing light hole and back to the gear legs. After takeoff, the 1/16 inch cable would be given a yank, which would pull pins to release the cable bridle between the mains. Then all of it would be reeled into the cockpit for stowage for the duration of the flight. It sounds like a page from Rube Goldberg—but it worked like a charm and added negligible weight.

Very heavy duty industrial rib 6-ply Goodyear tires were installed on the mains, inflated to nearly 100 pounds each!

The same auxiliary oil supply system that had been used on N7EZ in 1975 was installed on 79RA. It contained .8 qt. of oil and could be pumped via a hose into the crankcase in flight.

A heat muff was fitted around an exhaust stack with a hose extending up to the front cockpit. This was intended only as a backup source of cabin heat at the project cruising altitude of 11,500 feet where a nearly constant 0°C temperature was expected.

Fig. 7-5. Roger Houghton of the RAF holds the 74-gallon fiberglass back-seat fuel tank used on the record flight. Total fuel aboard at takeoff was 143.6 gallons.

Heavy clothing was intended as the primary means by which Dick would keep warm. The heater on/off "valve" was a rag stuffed in the end of the hose.

A glare shield was installed to prevent a fishbowl effect from the instrument lights on the canopy at night.

And at the 11th hour, a compression check on the 1500+ hour Lycoming brought the heartbreaking news that two cylinders were below acceptable standards. This resulted in one of the world's fastest top overhauls. Two new Slick mags were also purchased, however, one was found to be defective. Out of desperation as much as anything else, a replacement was dug out of RAF's junk pile, checked out and found to be in working order, installed and has been working to perfection ever since.

Two test flights were made, one during the day and the other at night, to check all the new systems—including the landing gear bridle release. Everything worked, so the aircraft was pushed in the hangar for one last exhaustive stem to stern inspection. The day before the flight, Friday the 14th, was spent attending to all the last minute details: getting the NAA observers briefed and sent to their stations, assembling the required clothing, food, extra batteries, and so forth, that would be needed.

The Long-EZ was impounded about 7:00 p.m. on Friday evening, and Dick hit the sack at 9:30. He left a wake-up call for 4:30 a.m.

Before dawn the crew, led by Burt Rutan, of course, converged on the RAF hangar and began setting the next two days' events in motion. N79RA was rolled out, started and warmed up, then taxied to the end of Runway 12. There, the tanks were carefully topped with STP treated gasoline and sealed.

As planned, Dick showed up at the last minute—ostensibly having had the chance to sleep right up to strap-in time. As the crew would later learn, however, he had been up for some time and had done his usual jogging before heading for the airport.

Donning gloves and an arctic coat over his other clothing, which included thermal underwear, Dick had himself literally stuffed into the cockpit. The airplane was gingerly pushed up ramps to the scales and was given its official weigh in . . . 1904 pounds for everything down to the GatorAde chewing gum. Well within the 2204 pound FAI Class C-lb limit.

The all-important barograph had been smoked and sealed, wound and set ticking. The Lycoming was started at 7:15 and the lift-off was marked by the NAA observer at precisely 7:27 a.m. A lot of breaths were being held as the Long-EZ hammered down the runway. Everyone who knows anything about flying realizes what an ultra-dangerous calculated risk a pilot is taking in an overgross takeoff such as this.

Dick was amazed at the Long-EZ's willingness to fly. Without hesitation the little beast settled into a 600 to 700-fpm climb and held it as Dick arced around to the right to pass the start/finish pylon to officially begin the record run. This done, he assumed his heading toward Bishop—and began reeling in the gear leg bridle. The retrieval was accomplished on what was a climbing downwind, just in case it somehow got back into the propeller.

Burt was pacing Dick in a Grumman Tiger and came in tight to watch and videotape the bridle retrieval. He accompanied Dick, in fact, for the entire first lap to insure everything was proceeding according to plan. Frequently Burt would slide under Dick to examine the belly of the Long-EZ to look for tell-tale streaks of oil or venting of fuel.

Depending on what sort of rate of climb the airplane had left after lugging its fuel load off the runway, Dick had had the option of flying up and down the Owens Valley at low altitude until enough gasoline was burned off to permit his climbing to a more efficient cruise level. However, with the EZ climbing so well, he chose to streak right on up to 11,500′ where he planned to spend the next day and a half. The Owens Valley is a very narrow and very deep fold in the tallest and most rugged mountains in the "lower 48." To the east rise peaks of over 11,000 feet and to the west is the very backbone of the Sierra Nevada. On each leg of every lap Dick would cruise past...and below...the 14,494-foot crest of Mt. Whitney. This is the valley that attracts sailplane pilots from the world over to test their skill and daring against what are believed to be some of the most powerful mountain wave conditions on earth, but the air was calm.

After the chase plane departed at the end of the first lap, Dick settled into a routine that consisted of a lot of chatter with the ground observers at each end of the course and a lot of fuel flow/leaning experimentation at mid course where he was out of radio contact. He soon found the particular EGT system installed in the Long-EZ was virtually useless as an aid in leaning. What worked was a little routine in which he established the fuel flow using a stopwatch and the Sears fuel counter, then matched this against various combinations of throttle and mixture settings. He found he used the least amount of fuel when he set up a constant rpm and then leaned until he got a 25 rpm drop. Any more or less used more fuel. Relaying the numbers back to Burt who ran them through his computer, it was soon evident that a very long flight was in reach. It would take 10 laps to break the record and they were shooting for at least 13—14 if things were going really well. Now, however, came the exciting news that if he could maintain the present fuel consumption, a full 15 laps might be possible.

Generally, the first day went well. The form-fitting, semi-supine seat of the Long-EZ was extremely comfortable with no pressure points developing. The engine continued to run smoothly and Dick was able to pass the hours with his fuel computations, log entries, checkpoint chatter and admiring the scenery.

"I swore up and down during the night that the stars were stuck in the sky," he recalls.

On one occasion while fishing down in a leg well for something he could keep on his stomach, Dick inadvertently bumped the mixture control and shut down the engine. The sudden silence instantly jerked him into a state of alertness.

"I looked and saw I had fuel and oil pressure, and for the life of me I couldn't figure what was wrong. The ol' adrenalin was flowing again, though, and after scrounging around the cockpit for a few seconds I discovered I had kicked off the mixture. Actually, it was a godsend because I was really fighting fatigue at the time. It gave me about an hour and a half of fatigue-free flight, which really helped.

"Having the moon come up about an hour before sunrise was a big event. I was very happy I got myself up for it and I was okay for a while. Then the sun came up and that was a big event—that lasted two or three hours. I felt good again, I felt normal."

About midnight, Dick had pressed on the left toe brake and had it drop to the floor. The next time he overflew Mojave, he advised Burt that he should be devising some way to retrieve a Long-EZ without a left brake. (The aircraft is steered on the ground with the brakes.) Burt said he would work something out.

In the early afternoon the decision was made—go for Lap 15. It would be tight, but more of a race with the sun than with dwindling fuel. Dick did not want to land without brakes after dark. If he could keep his speed up, however, he should touch down just after sunset but with adequate twilight remaining.

On the next to last lap, the Defiant joined up and escorted Dick the rest of the way. At long last he made his final turn at Bishop and headed down the home stretch.

FÉDÉRATION AÉRONAUTIQUE INTERNATIONALE

Diplôme de Record

NOUS SOUSSIGNÉS CERTIFIONS QUE *Richard G. Rutan* (États-Unis d'Amérique)
SUR *Long EZ 001*
A ÉTABLI LE, *15 et 16 Décembre 1979*
LE RECORD SUIVANT *du monde par catégorie: Distance en circuit fermé:*
Mojave/Bishop, Californie   *7.725,30 Km.*
Classe C·1·b Groupe 1

POUR N.A.A.
LE PRÉSIDENT   LE DIRECTEUR GÉNÉRAL DE LA F.A.I.   LE PRÉSIDENT DE LA F.A.I.

Fig. 7-6. Certification by the Federation Aeronautical International (FAI) of the Long-EZ distance record. This plaque is on the wall of Rutan's front office at Mojave.

"About halfway on the last leg, I climbed up to 13,500′ and when we determined we had plenty of fuel, I pushed the throttle wide open and set up a 200 feet per minute rate of descent. The airspeed pushed up to 180 or so, coming down hill all the way. After the last ridgeline, I got it right down on the deck. Everyone was waiting down there and I wanted to end it with something special — it was a little treat I had been saving for myself. I was really feeling good as I came across the field at about Warp 3, and once past the pylon I did a steep pull-up and roll-off and came around to set up for a landing."

The landing, of course, was with an emergency. Dick slowed the aircraft as much as possible on approach and after touchdown held the nose high to get maximum aerodynamic braking. As the Long-EZ decelerated to the minimum speed that rudder control was still available, he came in with just enough power to maintain that speed. At that point, a motorcycle ridden by Burt and with Mike Melvill seated behind him came roaring up on the left wing. After neatly synchronizing their speed with the Long-EZ, Mike reached over and took a firm grip on the left winglet. Seeing this, Dick cut the power and allowed Burt and Mike to quickly guide him to a stop—right in the middle of the runway.

Dick sat there for a second listening for the barograph, then gave a yell when he could hear it still tick-ticking away.

"I knew then we had all the squares filled, that we had a record that could be certified!" (Figs. 7-6 and 7-7)

FROM THE OWNER'S MANUAL

A complete 66-page Owner's Manual was prepared for the Long-EZ, which contains several comments unique to this design. The following are some examples.

Engine Start. Engine starting may be accomplished by hand-propping. While you have undoubtedly been horrified by the accident statistics on hand-starting antique aircraft, remember that the Long-EZ is a totally different story. Antiques are generally tractor aircraft, which means that they tend to chase you, once started. Long-EZs, on the other hand, try to run away from you. The traditional hand-start airplane has to be chained down and main wheels blocked for marginal safety (the tractor prop still tries to suck you in). The Long-EZ with nose-down parking chocks itself, and the pusher prop blows you away from danger. With modern, impulse-coupled magnetos, it is not necessary (or desirable) to make a Herculean pull of the propeller for starting; just pull the engine up on compression and give it an EZ flip through. In the unlikely event that your Long-EZ does run away from you after starting (if you leave the throttle open), it won't carve the first thing it comes to into hamburger, but will give it a bump with the nose instead.

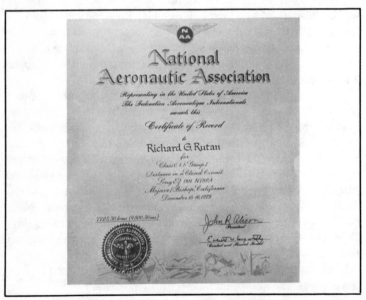

Fig. 7-7. Official recognition of the Long-EZ record was also made by the National Aeronautic Association. The plaque is also on display in Mojave.

Fig. 7-8. Large 500 × 5 tires are tested on the Long-EZ as Burt Rutan takes documentary footage with a tape movie camera. The 2 × 4 plank under the aircraft was used to simulate a rough field before the aircraft was actually taken into a small nearby airport.

Rough Field Caution. Although the Long-EZ may use the larger 500 × 5 tires, this does not make the aircraft totally suitable for rough, gravel or unprepared fields (Fig. 7-8). Since the Long-EZ is a pusher, the aircraft cannot be rotated as easily as a conventional tractor aircraft. You still must accelerate to normal rotation speed, 50-60 knots, depending on cg, before the nose wheel comes off, and during this time the nose wheel can kick debris into the prop. The small nose wheel tire, high rotation speed and prop damage possibility make the Long-EZ less suitable for unprepared field operation than a conventional aircraft."

A standard non-turbocharged Long-EZ (N79RA) attained an altitude of 26,900 feet in December '79. CAUTION: The altitude capability of this aircraft far exceeds the physiological capability of the pilot. Use oxygen above 12,500 feet.

Caution-In the Rain. When entering visible moisture (rain) the Long-EZ may experience a pitch trim change. The Long-EZ prototype (N79RA) has a significant nose-down pitch trim change in rain. The VariEze prototype has a mild nose-up trim change. Some VariEze owners report nose up and some nose down. This phenomenon is not fully understood and your aircraft may react differently. Our flight tests on the prototype Long-EZ have found a slight performance loss and the pitch trim change forces could be trimmed hands off with the cockpit trim handle at airspeeds above 90 knots when entering rain. Once the aircraft is in visible moisture conditions, it can be retrimmed and flown normally. There may be a disorientation factor during the transition from visual to instrument flight that the pilot must be ready for, especially if your trim change is significant. If your rain trim change is found to be significant, install a placard to notify pilots of this characteristic.

Landing. Make your approach and traffic pattern very cautiously (Fig. 7-9). Most pilots and controllers are accustomed to looking for more conventional aircraft of gargantuan proportions (like Cessna 150's) and may ignore you completely. Best pattern speed is 70-75 knots (80-85 mph), slowing to 65 knots (75 mph) on final approach (70-75 knots in turbulence or gusty winds). The Long-EZ is a very clean airplane and you can double the runway length required if you are 10 to 15 knots fast on your approach.

Caution—Parking. With the nose gear extended and without the pilot in the front cockpit, the Long-EZ may fall on its tail. The aircraft may initially sit on the nose wheel, but may tip backwards when the fuel bleeds through the baffles towards the aft of the tank. Be sure to brief all ground handlers that the aircraft can fall on its tail unless parked nose down and could also get away from them while moving the aircraft. If your aircraft is subject to being moved by unknowledgeable people, ballast the nose or attach a sign to caution them about the possibility of tipping over."

Pilot Experience Requirements—Pilot Checkout. There is no such thing as a minimum number of total hours a pilot should have to be qualified for checkout solo in a new aircraft. The best pilot qualification is variety. He should be current in more than one type of airplane. The Long-EZ is not difficult to fly, but it is different; like a Yankee is different from a Cessna, or a Cub is different from a Cherokee. A pilot who is used to the differences between a Cessna and a Cub is ready to adapt to the differences in a

Fig. 7-9. Long-EZ flown by Dick Rutan on a landing approach at Mojave. Note temporary camera mount fitted into the nose of the aircraft.

Fig. 7-10. Sally Melvill was the first woman to solo the Long-EZ. She had 150 hours at the time.

Long-EZ. The Long-EZ has entirely conventional flying qualities. However, its responsiveness is quicker and its landing speed is faster than most light training aircraft. It should not be considered as a training airplane to develop basic flight proficiency.

ANOTHER HEADLINE?

While a two-airplane flight, non-stop—one starting from Cape Canavaral and the other from San Francisco—to the mecca of homebuilders, Oshkosh, isn't too challenging, it probably hasn't been done before. Thus, Johnny Murphy, Mayor of Cape Canaveral and the sixth VariEze builder to fly, teamed up with Dick Rutan just as soon as his second-to-fly Long-EZ was completed, to plan the above flight. He would take off with his son from Florida and fly non-stop to Oshkosh. Dick Rutan would take off from San Francisco and do the same thing. By prior arrangement, the pair plan to arrive at the EAA fly-in together. We'll have to wait and see.

LADIES' FIRSTS

The first woman to solo the Long-EZ was Sally Melvill (Fig. 7-10). She was also the first lady to solo the VariViggen. Between answering the phone at the RAF and opening the day's mail, Sally expressed her feelings with the Long-EZ checkout this way.

"It was all so easy! I had approximately 150 hours when I soloed the Long-EZ. Dick Rutan gave me two check rides and turned me loose. The first impression I had was a feeling of being a part of the airplane, and of wanting to dance in the sky. The Long-EZ needs very little hands-on flying—it seems that you think of what you want and where you want to go, and it does it. I had had the feeling that it would be a sensitive airplane to fly, but I found that if you did over-control, just let it go! The airplane seemed to know how to fly better than I. It is not a difficult plane to fly—just different. Sweet is a word that fits for me. A great feeling of being a part of the airplane. I enjoyed the Long-EZ so much that Michael and I are busy building one for ourselves. It is going to be hard to keep me on the ground. Long-EZ—WOW!!

The lowest time pilot to solo the Long-EZ to date has been Patricia (Pat) Storch (Fig. 7-11). Not counting passenger time in other Eze's, she had 24 1/2 hours of dual and solo before Dick Rutan gave her an hour and a half of dual instruction. Thus, she had logged just 26 hours flying time when she soloed the Long-EZ at Mojave. Here's what Pat had to say.

Fig. 7-11. Pat Storch had only 26 hours flying time when Dick Rutan checked her out in the Long-EZ. However, she had logged many hours of passenger time in the canard aircraft.

208

Fig. 7-12. Designer Burt Rutan flies his Long-EZ in formation over the Mojave Desert. Burt, a 2,000-hour pilot, flies an excellent formation.

"Incredulous—that was my first feeling when they told me they wanted me to solo the Long-EZ. Tiny insecurities worked their way out in the form of protests; 'But I'm only a student! I've only soloed one other airplane! I have less than 30 hours!' It seemed that I was the only one lacking in confidence, because they would not be dissuaded.

"The day came when it was time to give it a try from the front seat. The cockpit looked foreign, almost hostile. Instruments were not where my eyes wanted them to be. Throttle and stick were in the wrong hands. With my heart in my mouth, we started the pattern work. Soon I was thankfully too busy to be nervous, but I still felt I was reaching for an unattainable goal. Control of the Long-EZ felt so different, and the full stall landings I had practiced so diligently in the Tiger were to be forgotten.

"Then, amazingly, little pieces started falling together. Each landing felt better; the cockpit looked familiar, and a tiny seed of confidence started to bloom. Could it be? Would it really happen? Down to refuel and then came the words I wanted to hear: 'You're ready to go!'

"I was full of anticipation and busting with excitement. The takeoff was smooth and felt good; the plane felt fantastic. Then I wanted to play in the sky—up, down, turns and steep turns.

"I never expected an experience to equal my first solo, but this surely surpassed it. Flying never felt so good. Then came the final test—the landing. A little long, but a good one.

"A Long-EZ pilot! I may have landed, but I was still in the air. In fact, I haven't come down yet. It was the most satisfying exhilarating experience I've ever had."

What better way to prove the docile flying characteristics of the Long-EZ (Fig. 7-12) than to kick a chick out of the nest, and do it safely, with just 26 hours of flying time?

WE FLY THE LONG-EZ

You can do all the research on a project, study its history, talk with builders and share their experiences; however, there's nothing like flying the product to make all this information fall into place. That's what we did on a crystal clear day at Mojave with Instructor Dick Rutan sitting lightly in the back seat of the No. 1 Long-EZ, N79RA (Fig. 7-13).

None of the VariEze series has been designed for training purposes and the "stretched" version is no exception. In the bare back seat, Dick had a side control stick and that was all. There were no rudder pedals and brakes, no throttle, no mixture control, no radio and no intercom. Dick Rutan checked his log book on 79RA and noted that he had "checked out" 24 other pilots since original flight testing and the 33-hour, 34-minute record flight.

"Everyone we put in the front seat has to be a well-qualified pilot first. Most of our pilots are Eze builders who have flown their own ships extensively and really have nothing new to learn," explained Dick. "I've given many of these Eze pilots just a systems check and sent them out in the Long-EZ. Nobody has had any

Fig. 7-13. "Here's the way you do it," explained instructor Dick Rutan, talking with his hands, as he gives a pre-flight briefing to Don Downie (courtesy Plane & Pilot by Bob Weston).

Fig. 7-14. Instrument panel on the Long-EZ. Essential blind flight instruments are installed, along with a single nav/com radio.

trouble. Some pilots who have never been in a canard before will tend to level off too high, but we have plenty of runway here at Mojave."

Yes, there was plenty of runway, and I used most of it! We'll explain why later on.

The Long-EZ cockpit is quite comfortable once you begin to get accustomed to the supine position. The cockpit, two inches wider than the VariEze, is wide enough and adjustable rudder pedals will give sufficient leg room. I'm a long-legged 6'2" and we should have taken the time to extend the rudders for my legs. Instead, we removed some seat cushions and went on from there. My shin bones and the bottom of the instrument panel had a speaking relationship by the time we had landed.

Dick Rutan has been a flight instructor for nearly 20 years, and he does an excellent job—both in pre-flight briefing and later in the air. He helped me unfold inside the cockpit and get the seat belt and shoulder harness adjusted. Then came a systems check of all the goodies in the cockpit—and it is quite "busy." Starting left to right, we found the canopy latch and emergency locking system, a firm reminder that it is extremely poor policy to take off with the canopy not fully down and locked.

Both the canopy lock and gear down system are wired into a warning horn and light in the middle of the panel. A small switch next to the throttle will disarm the horn, but not the light, for throttle-off, nose-gear-up practice.

The instrument panel is straightforward (Fig. 7-14). The

super-dependable Lycoming 0-235 has an adjustment for carburetor heat, but Rutan advised that it wasn't necessary in the dry desert climate.

A typewritten checklist taped near the pilot's left elbow reads as follows:

Before Take-off:
1. Fuel caps/oil cap locked
2. Controls free and correct
3. Seat belts fastened
4. Trim for take-off
5. Circuit breakers IN
6. Dyno ON
7. Lights as required
8. Fuel - fullest tank
9. Master ON
10. Boost pump ON
11. Start - left mag only.
12. Ground run-up
    Mags 2000 rpm
    Oil pressure - green
    Fuel pressure - green
    Dyno - load check
13. Canopy locked

Climb-Cruise
1. Gear up
2. Boost pump off
3. Mixture lean as required

Before Landing
1. Fuel - fullest tank
2. Circuit breakers IN
3. Mixture RICH
4. Boost pump ON
5. Landing gear down - indicator check

After Landing - Engine Shut Down
1. Boost pump OFF
2. Lights OFF
3. Dyno OFF
4. Radio OFF
5. Master OFF
6. Mags OFF
7. Record tach time

The "dyno" used for electrical power on the Long-EZ is a four-pound unit from a Kobota tractor that produces unregulated alternating current. A standard alternator weighs between 15 and 20 pounds.

Not listed on the checklist, but used in practice, is to plug in the headset and microphone after the pilot is seated and remove it before deplaning because the plugs were installed between the pilot's legs. "We'll change that location later," explained Dick.

One other cockpit placard lists "Front Seat Pilot weights; Maximum 250 pounds; Minimum 120 pounds."

Beside the pilot's right elbow was a hand-calculator-type "Long Ezcale" being developed for functions of TAS, GPH, Fuel used/remaining, time, battery voltage, inside/outside temperatures, and other cockpit calculation functions. The Rutans plan further development of the "Ezcale" with a view to marketing in the near future.

Soon it was time to flick on the right mag switch and Dick walked around the Long-EZ and cranked up the sturdy 115-hp Lycoming (Fig. 7-15). I had the throttle cracked a bit too far and it took a few pulls on the prop to get ignition. Then the other mag switch came on and Dick walked to the front of the ship where he lifted the nose easily and I cranked down the nose gear with about 10 turns of the easy-to-reach knob in the middle of the instrument panel.

We kept the radio on Mojave's 122.8 unicom and handled in-flight instruction just by voice. The pusher's cabin was surpris-

Fig. 7-15. Dick pulls the prop through as "student pilot" Downie sits up front (courtesy Bob Weston).

ingly quiet so that an intercom headset system was really not required.

After having read voluminous material on the non-standard "side stick" control, I'd expected some problems. However, by keeping your right elbow snugly in the arm rest, there was little tendency to overcontrol.

Rutan slid up on the leading edge of the wing and told me to taxi out. The rudder and brake attachments take some getting accustomed to, particularly when you wear a big size 12 shoe.

The nose gear casters so that all steering during taxi is done with the brakes. Pilots who have spent considerable time with the American Grumman series of trainers and four-place Grumman Tigers will find this sort of transition fairly easy. To others, there is a learning curve.

The sun was warm, but there was only a three-knot wind reported as we taxied along the rough ramp toward Runway 7. Dick remained on the wing (Fig. 7-16) until we neared the run-up area and then eased his frame into the back pit of the Long EZ. This was the location of the 74-gallon auxiliary fuel tank used on the initial record-breaking 33-hour, non-refueling distance hop.

Runup really doesn't amount to much: fuel boost as a back-up for the engine drive pump, a mag check, controls free, and you watch the trailing edge of the canard move in front of you as you pull back on the side stick. Then carefully close and lock the canopy. There have been fatalities as the result of VariEze take-offs with the canopy unlocked, even though the aircraft is fully controllable with the canopy open.

We called Unicom, were given a clear shot and taxied into position. As the power eased on, it was a bit of a guessing game to keep off the brakes and try to get a feel for when the tiny rudders were useful. With a high-performance cruise prop (58 × 72), acceleration was not extremely rapid, but our altitude was 2787 and the comfortable 75° day put our density altitude at 4472 feet.

We used up considerably more of Mojave's runway on take off than was really necessary because of our unfamiliarity with the novel rudder and brake system on the Long-EZ. On a subsequent solo flight, take off was in just about 1/3rd the distance and reasonably close to the 550 feet shown in the specs.

In any event, the rudder/brake pedals on the Long-EZ are completely independent of each other. Push in left rudder and you get just that; push harder and you get both left rudder and left brake. Push easily on both rudder pedals and you have both winglet

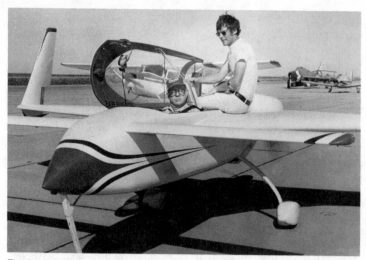

Fig. 7-16. Dick Rutan rides the cool back seat of the Long-EZ as Downie taxies in. Flying the Long-EZ is really quite an experience.

rudders deflected. Dick uses this procedure in lieu of flaps or a dive brake for rapid let-downs. Tense up on touchdown and push hard on both pedals and you lock both brakes—and probably blow both mainwheel tires.

On takeoff, I had some problems with directional control and snubbed the brakes a couple of times to keep the nose going down the runway. This, of course, extended our take-off roll.

As soon as the rudders began to take over and I eased my feet off the brakes, we picked up speed more rapidly. With the stick full back, the nose started to pick up as the airspeed went through 50 mph. The Long-EZ flew itself off the runway smoothly and I eased off the back pressure to go to a 90-mph climb speed. As soon as we were over the far end of the runway, Dick advised me to turn off the fuel boost pump and crank up the nose gear.

Our rate of climb with full throttle was close to 1000 fpm at 1300 pounds gross weight and density altitude as we circled out away from the airport. The control touch of the Long-EZ was smooth and comfortable. The supine position—actually your back leans 42° from vertical in level flight—would probably become completely comfortable in time.

There is something completely different about flying behind the stubby canard airfoil. After years of boring holes through the sky, you become accustomed to a conventional engine cowling or, in larger twins, vast nothingness in front of the windshield. But to have that stubby wing breaking the air ahead is quite different. Even piloting an open biplane from the back seat where wing structure is ahead and above you has little bearing on the view from the front seat of the Long-EZ. Visibility is excellent.

On this first flight, I felt that I had to carry a little "bottom rudder" in turns to keep the ball in the center. We discussed this on the ground with Dick during the two-week period between two flights. Dick advised that his tests found only a slight amount of adverse yaw with aileron use and almost no torque or "P" effect during the slight pitching moment found with power changes. But more about these fine points during the second flight.

We leveled off at 7500 feet and Dick asked me to lean the engine to best power (maximum rpm) on the fixed-pitch prop. We went back, and back, and back some more on the mixture—almost to the idle cut-off position—before the little Lycoming began to roughen up. Then forward to a smooth engine and a try for maximum cruise speed. Full throttle at 7500 rpm with a non-turbocharged engine equals 75 percent power and is permitted indefinitely on this engine.

The airspeed picked up to 163 mph or 164 calibrated. Compensate for altitude/temperature, and we were doing 187 mph—and on 75 percent of 115 (or 86 hp) this is *really* moving. This same engine will drive a conventional two-place Citabria at 116 mph. The Long-EZ is clean-clean!

We throttled back to a comfortable cruise of 19 to 20 inches (60 hp) and tried some turns. Visibility is excellent and all the controls were most responsive without being "touchy." Rate of roll was excellent. In medium-to-steep turns, the canard up front made an easy reference for both pitch and turn angles. The long main wing with its tiny rudders at the tips took some rudder going into a turn, much like a long-spanned sailplane.

After a high-speed run, Dick suggested sampling the slow speed end of the dial. We cut back to 15 inches and then to 12 inches to let the engine cool slowly. Then the power came all the way off and I eased the pitch trim lever full aft. The air speed dropped into the "approach" segment of 70-80 mph and it felt so slow that I thought we were about to fall out of the air. Dick has fixed up an idiot-proof overlay on the airspeed indicator. He has marked cruise climb in the 100-110-mph range, best rate of climb at 87, pattern speed at 80 mph, and final approach at 72-75 mph.

"Now bring the stick all the way back and see what happens," counseled Rutan.

I eased back on the stubby side stick until it hit the aft stops. The 11.8-foot canard came up slightly and our airspeed finally dropped to 60 mph. There was that mushy feeling of an impending stall—but perhaps all in my mind. The nose dropped slowly, perhaps a degree or two. The Long-EZ picked up perhaps 2 mph, and we were right back flying again.

Dick then asked for more turns. "Make them as steep as you wish with the stick all the way back."

I tried turns up to 70-80° and we still didn't stall—though our rate of descent increased dramatically. The rate-of-climb needle dropped to 1,000-1,200 fpm before roll out, but there was no tendency for the ship to duck a wing or roll inverted.

We tried stalls and promptly found that you just don't stall one of Rutan's canards. We tried accelerated stalls and the Long-EZ would merely "nod" its canard slightly when the front wing "stalled" and with an increase of a mile or two an hour, the wing promptly unstalled. You can fly around all afternoon, turns or not, with the stick all the way back and all that will happen is that you'll develop a mild rate of sink, perhaps 600 fpm, with power off. At 15 percent power, you hold level flight with the stick full aft. That's your practice for the pattern and landing.

"We have this designed so you can't force a 'departure,'" explained Rutan. A "departure" is departing from controlled flight and can result in a Lomcevac maneuver where the airplane will tumble completely uncontrollably until it makes up its own mind to start flying again.

Then came steep climbs, full stick back, both straight ahead and in turns. With the supine seating, it felt that you were going straight up, but a check out the side indicated a 45-55° angle of climb. As the airspeed dropped, stick full back, the nose would drop slightly all on its own and the Long EZ would be back flying again.

It was all most impressive!

"This trick will help get you killed in a conventional airplane," grinned Dick from the back seat. And he was quite correct.

The air over Mojave was busy on a clear weekend. A gaily painted AT-6 and a surplus British Vampire jet fighter were rolling around in the sky near the airport. Mojave is both a sport plane center and near the Edwards AFB restricted test area, so many unusual things occur in this airspace. And we had more than our share of non-standard operations that afternoon.

Dick pointed out the jet as it circled off to the north of us. As we watched, the jet began a slow left turn that seemed destined to intersect our flight path. We dumped the nose on the Long EZ to get out of the way and the fighter dropped his nose slightly. Dick swore quickly and we kept going down. However, our 185 mph was no match for the turn rate of the jet, and we had a far-too-close look at the air intakes as he flashed just behind our twin tails. At that time, Dick didn't know whether the jet was "playing with us" or if the other pilot never saw us. In either event, it was too close.

In the quiet cockpit of the Long-EZ, Dick and I exchanged irreverent and unexpurgated comments to unwind our nerves a little. Then we circled back to make a landing.

On a previous "check ride," I had been at the end of the runway shooting pictures and watched a VariEze owner make three passes at Mojave to get the Long-EZ slowed down enough to "make the field." At the time of this flight, the Long-EZ did not have a speed brake. The plans have already been drawn and all Long-EZs will have an under-the-fuselage speed brake.

We cranked down the nose wheel and could see it extend through the small plexiglass window in the cabin floor. The boost pump went on and I punched the transmit switch on the EdoAir radio to check traffic at Mojave Unicom. There wasn't anything reported, but we were still very wary of a T-6 and an old jet fighter. Once was enough on that routine!

Because the EZ is so clean, and without the speed brake, Dick recommended only a 500-foot traffic pattern. Had we flown the standard 800-foot circuit, we'd have never gotten to the ground. However, Unicom at Mojave was accustomed to this procedure. I chopped the power and waited and waited, all the time easing back on the stick so that we could slow to approach speed. It took the whole downwind leg. That miniscule 1.6 square feet of frontal area is certainly noticeable when you try to get this bird slowed down.

Over the railroad town of Mojave, we looked around and turned on a wide base leg. Visibility from the front seat is excellent in all attitudes except during landing flare. We rolled out slowly to line up with Runway 7. We were still a little fast according to Rutan's "idiot-proof" airspeed overlay, but Dick said, "You're in the groove; just stay with it."

We crossed the numbers at perhaps 20 feet and 65-67 mph. Then came the problem of slowing down. We eased back more on the side stick and the canard came up to cover our view of the far part of the runway.

Dick's voice came from the back. "You're still ten feet in the air; let her down a little."

I eased off the back pressure just a bit and we continued to eat up runway. It was a good thing that we had 7,000′ to work with because the little 3.40 × 5 main tires didn't touch until almost the midpoint.

"Now let the nose wheel down and begin to use some braking," advised Rutan. We followed instructions. We played with the brakes gingerly to avoid skidding the tires and let the ship roll to the far end of the runway.

As we turned off to taxi back, Dick asked for the canopy open so that he could stay cool. It seemed like a good idea, though the glass house was not really uncomfortable even on the ground.

We taxied back to Rutan's ramp and closed the mixture control. As the engine stopped, Dick slid off the wing and lifted the nose. I cranked the nose wheel up and he eased the stubby fuselage to the ground. It's certainly an unusual way to park an airplane—any airplane but Rutan's designs.

I snapped all the switches off, unhooked the harness and belt and slowly extricated my long legs from under the instrument panel. It was only a short step to the pavement.

As we re-hashed the flight, Dick handed me the log book of the Long-EZ, put the date and my name on a vacant line and said, "Tell me what you think of the flight." There wasn't much space, so I signed it and said, "Great ship!"

Fig. 7-17. A familiar sight at the EAA air shows. Rutan's Defiant and Long-EZ in formation with the VariViggen joining the group.

You'd think that was enough excitement for one day, but we decided to shoot some air-to-air formation pictures of both the Long-EZ and Mike Melvill's No. 1 VariViggen (Fig. 7-17). Dick took my flying buddy Bob Weston in the back seat of the Long-EZ and Les Faus, who had flown his VariEze up from Van Nuys, went along in our Cessna 170 (Fig. 7-18) to do the flying while I shot pictures.

"I'm not current in taildraggers," said Les as we taxied out, "so you please make the take off and landing." That's the way we did it.

We took off first with the two canards right behind us. As the Long-EZ "formed up" with the Viggen and climbed toward our camera platform, Mike came up on the 122.95 air-to-air frequency we were using. "Did you see that F-4 go past us and then go right straight up?"Dick Rutan acknowledged and back-seat passenger Bob Weston told later of seeing the supersonic jet go past the two canards and pull straight up out of sight. But more on that in a minute.

According to plan, Mike came in alone first for a load of 11 black-and-white shots from my 6×7 Pentax (200 mm Takumar lens). Mike flies a great formation, and we had to keep admonishing him to move away a little so that we'd get the entire airplane in the photo.

Then we changed film and Dick moved in behind Mike with the Long-EZ (Fig. 7-19). We shot another 11 pictures and a few assorted color shots with a second camera. As Mike pulled away, we noted a large black ball of smoke erupting from the desert floor north of Mojave toward California City. Since Mike's camera posing was finished, he headed for the smoke while Dick and his passenger came in for solo pictures. We soon finished the roll and Dick headed toward the fire that was billowing from the desert floor.

"That looks like a plane crash," said Mike on our plane-to-plane frequency. "Let's change over to 'point eight' (122.8) and advise the airport. Dick and I "rogered" and we all went to the busy Unicom frequency. We followed the two canards toward the fire and I could hear Mike talking with Mojave's Airport office. He confirmed that the smoke was a burning aircraft.

"It looks like an F-4," said Mike. "It's probably out of Edward [AFB]."

Mojave Unicom, with Airport Manager Dan Sabovitch behind the radio, said, "Please change over to 120.7 and advise Edwards

Fig. 7-18. Here's how these air-to-air photos were taken. A back-seat photo of the author's Cessna 170B shows a portion of a canard wing in the foreground. Air-to-Air photos were taken with a 6 × 7 cm Pentax and an F.4 200m Takumer lens.

Tower of what you see."

About that time, Mike came on the air from the front seat of the Viggin. "Hey, there's one parachute coming down, but he's still way above me." Sure enough, we could see the red and white 'chute drifting lazily toward the calm desert floor. Below, streams of dust converged on the burning F-4 and the probable touchdown of the survivor. I don't know where all the motorcycles, campers, dune buggies and standard cars came from.

Fig. 7-19. Dick Rutan enjoying himself flying formation with the Long-EZ.

Later, Dick Rutan commented wryly, "It was just like North Vietnam all over again. Whenever a 'chute came down, there were people crawling out from under rocks headed for the impact site."

The airwaves were busy with queries about a second parachute since the F-4 always flies with a two-man crew. There was no second 'chute.

Yes, it had been an unusual day.

The first flight of the Long-EZ was in mid-1979. Final flight testing was accomplished in December and the initial record flown December 15-16. After the successful record flight, the first passenger in the Long-EZ was Burt and Dick's mother, Irene. Since that time, the popular ship has been giving demonstrations and flight checks. Dick had just returned from Florida's annual Fun'-n-Sun fly-in at Lakeland where he flew a waivered airshow in the Long EZ for each of the seven days of the program. He logged 52.2 hours on the two-week trip. Flight time to Florida from Mojave: 4:20 to El Paso and 8:10 non-stop into Lakeland. Average true airspeed was 183 mph on 5.5 gph. Altitudes ranged from 9,500 to 13,500 feet. The engine, now with 1534 hours total time (it was purchased used and put into the Long-EZ with 1350 hours when it was removed from an American Yankee trainer) required a main oil seal that was replaced in Dallas, Texas, on the way back. Aside from that, it was just gas and oil.

After the Long-EZ returned from Florida, we had the opportunity to fly N79RA again—this time without Dick's 170 pounds in the back seat (Fig. 7-20). On any initial orientation flight in a new style airplane, the pilot is somewhat overwhelmed with new operational procedures, new reference points, and a whole new cockpit arrangement. After the first flight, we had roughed-out the copy and had shown it to Dick.

"You'd better go out on your own and correct a few of your misconceptions," suggested Dick after reading our rough draft. He then suggested a mini-test program to prove or disprove to my own satisfaction some of my initial impressions. Before flight, the former USAF fighter pilot suggested that I pay particular attention to solo rate of climb, any yaw abnormality, adverse yaw with full aileron and feet on the floor, torque, "P" factor pitch trim changes with full power application, visibility at various climb and approach speeds and simulation of a "Hormel-handed" pilot who might tighten up his turn in the traffic pattern when a strong wind was blowing on the ground.

Rutan opened the nose compartment and readjusted the rud-

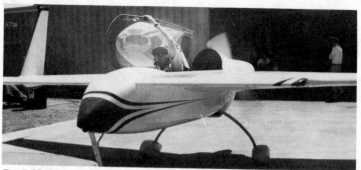
Fig. 7-20. From here on out, you're on your own. Downie taxies from the RAF for a solo flight in the Long-EZ.

der pedals to full extension. It was a great help to me. I shoehorned into the cockpit and slid the back cushion as far aft as possible. Then the paper listing the items to be checked in flight went under one leg and I followed the cockpit checklist. Dick propped the Long-EZ while the nose was still on the concrete. Then he and a friend lifted the nose and I cranked down the nose gear, squirmed in the seat, called Mojave Unicom and taxied out. The group of aficionados who regularly congregate at the Rutan hangar on Saturdays wasted considerable film, and undoubtedly even more envy. Flying the Long EZ solo is a real treat!

There had been a wind shift and noontime thermals were starting to pop. Unicom advised either Runway 22 or 30. I taxied out slowly, waiting for two aging British jets to exhaust their fuel with fast passes down the runway. Both finally landed as I completed my run-up. After a call to Unicom, I lined up on the spacious 9600-foot Runway 30 and eased in the power. I "worried" the rudders and applied brakes as required to keep the nose pointed more-or-less down the runway while easing in full power. Acceleration solo and with low fuel was much more rapid than on the previous dual flight, and there was less inadvertent brake dragging.

I hopped into the air, adjusted the trim, retracted the nose gear and made a slight left turn to parallel the parking ramp. With full power, the rate of climb swung between 1200 and 1500 fpm in moderately choppy air at 90-100 mph. Rate of climb is actually better at 80 mph, but forward visibility is poorer.

I shut down the fuel boost pump, came back on the power to about 22″ and climbed to 6,000 feet indicated. After leveling off, my first test was to take both feet off the rudders and roll the wings quickly, checking for adverse yaw. With an abrupt roll to the right, the nose tended to go up and to the left, but ever so slightly—not more than 4° or 5°.

Next I throttled back and let the speed bleed off, watching how visibility was affected as the nose and its canard came up. The higher the nose, the more obstruction to vision. At pattern speeds, the canard was clearly visible but presented no problem; on a straight-ahead, full-stick back stall, it was a slight obstruction. I let go of the controls with full aft trim, let the ship stabilize and briskly applied full power. The nose slowly dropped perhaps 5° and wandered off to the left very slightly. With power reduced, the nose returned to the horizon. Good, safe, docile.

While I was slowed down, I simulated the careless pilot trying to "hurry" a turn toward the airport. With full back stick and ample rudder, the Long-EZ turned calmly. Holding the stick back in a continuing turn, there was a mild oscillation as the canard stalled, lost just a bit of its lift and the nose dropped slightly. Just as soon as a little bit of speed was picked up, the ship was flying again smoothly.

I had taken off with only about 1/3 fuel and Dick had suggested that I return within 30 minutes. I called Unicom and was advised that Runway 22 was now in use. That's the shortest of Mojave's three runways, but it is 5200 feet long with nothing but sagebrush on approach.

I flew a long pattern, slowed to the 75 mph marked in the airspeed box for approach and held perhaps 70 indicated across the numbers. During flare out, it was still a bit hard to see around the canard, and I found myself taking a quick glance or two out to the side as I let down through about 10 feet. There was a minimum of float and the mains were on. Even without Dick's weight in the back seat, there was no holding the nose gear off with full back stick. It plopped to the runway and I let the Long-EZ roll out. It took several power applications to reach the end of the runway.

I cleared the active, opened the latch on the canopy and pushed it up to keep the desert heat tolerable. The taxi back was with one hand holding the canopy open.

Yes, the Long-EZ proved out all the questions I had from that initial demonstration flight.

Then Dick checked for oil leaks, added a little fuel and went out to work on his air show program while a dozen VariEze builders and would-be builders watched.

It's a great airplane!

Chapter 8
Everybody Gets Into The Act

There's much more to joining the homebuilt scene than sending your check for a set of plans, stopping at the neighborhood aircraft supply house, and moving one car out of the garage.

Homebuilding is a people-on-people experience. There's the designer working with the builder, the builder working with the designer, and builders working with other builders.

THE CANARD PUSHER

Faithfully every quarter, Rutan produces a mimeographed newsletter that goes to all builders of any of the various projects. This newsletter is a mandatory publication for active builders since it is the only way the homebuilt designer can communicate officially with any update on plans.

This newsletter is the primary clearing house for ideas, shortcuts, vendor information, fly-ins, and just about everything a builder wants to know.

Originally called the *VariViggen News* (the first six issues being concerned with the VariViggen exclusively), the name was changed with the advent of the VariEze to *The Canard Pusher*, in October, 1975. The newsletter name, which has continued for at least five years, was picked over a number of other interesting mastheads submitted by builders. These included: *The Canard Line, The Canardian Club, Vari New-Z, Canard Time News, The*

Vari Forum, VariUnique Aircraft News, Canard Capers, Canard Contrails, Glass Backwards News, VariVignette, The Canard Rumor, Canard Disclosure, Canard Gas Line, Canard Trend, Canard Courier, The Backward Flyer, The Canard Leader, On the Nose, Canard Tales and *Canard Forward.*

"The new name should last for some time now, until we come up with an aircraft without a canard," said Burt Rutan.

The evolution of the unique landing gear on the VariEze and Long-EZ is an example of how the designer helps supply hard-to-fabricate parts. This change was duly noted in the builder's newsletter.

The vendor for the VariEze fiberglass parts notified us that he was modifying his business operation and would discontinue manufacturing the nose and main gear parts. Since tooling for the main gear was wearing out, we would not arrange for the new manufacturer to produce this same item. Instead, we developed a new gear, which is 40 percent stronger and made with an improved process. This gear is designed specifically for the Long-EZ, which has a gross weight 24 percent heavier than VariEze. We plan to have only this new gear produced by the new vendor, and use it on both the Eze and the Long-EZ. It is three to four pounds heavier than an Eze gear. We have also developed an improved main gear attachment design which is now being tested on the Long-EZ."

After testing, Rutan advised that "The search for a manufacturer was difficult—the normal fiberglass production shops do not have the necessary ovens, instrumentation and impregnation

Fig. 8-1. Long-EZ landing gears are formed here at the RAF in Mojave. A special production method was developed by Rutan to wind filament and which uses a new epoxy system. The new gear is 40 percent to 60 percent stronger and 24 percent heavier than the original.

Fig. 8-2. Roger Houghton places new Long-EZ gear legs in a special convection oven for curing. This system was developed by Rutan when vendor bids were considered too high.

machines needed for this work. The shops that do have the equipment and engineering capability to produce these parts are those who only make expensive aircraft components and they have bid excessive prices on these items to cover the development costs. We were faced with either no landing gear, or a price increase.

"We felt that the price of the gear, having been raised 20% in the three years, so excessive, we decided to tackle the job ourselves—the only alternative available to provide a reasonable gear for the builders. We knew this job would result in delays in plans preparation for Long-EZ, but it really was our only alternative. Within six weeks we produced patterns and tooling for the new strengthened nose and main gear (Fig. 8-1), developed a new production method, built a convection oven (Fig. 8-2) and filament winding machine, set up a quality control process, tested a new epoxy system, and began production of the new parts. Our new method, which winds roving and tapers it for accurate mold filling, requires fewer man-hours and results in more uniformity than the previous process. Thus, we are able to sell the parts for only 70 percent of the previous price per pound.

"The new main gear is 65 percent stronger than the old VariEze gear and should not be susceptible to the long-term creep (spreading and loss of camber) experienced on some of the heavier VariEzes. The new gear is 3 inches longer on each end to raise the prop clearance on the Long-EZ. VariEze builders using the new gear will saw off the 3 inches. The old main gear weighed 16 pounds. The new one weighs 21 pounds. While we regret any increase in empty weight, experience has shown that many VariEzes are being operated heavier than expected and need the extra beef in the gear."

Contributions to the newsletter by the builders indicates a wide spread in the age spectrum. It was reported that VariEze N9036G was built by George Gilmer, age 73 years young. Three years total time at Santa Paula Airport, California. Total cost was about $3600. George made almost every part, wing fittings, wheel and brake system, throttle quadrant, etc. George advised all builders that if Burt gives a measurement or method, there is a reason. George started flying in 1929.

On the other end of the age spectrum is this report on a first flight by Stuart Kingman, of Pacino, California, age 18.

"After three years of blood, sweat, and tears, I had my own airplane! I taxi-tested for two or three days just to get the feel of the airplane, and also because I was scared, and rightfully so after reading of all those experienced pilots getting killed in their Ezes. All of my experience rested on 150 hours in a Cessna 150 and one hour in a Grumman TR-2. After the taxi tests, my Dad took out a $10,000 life insurance policy for me, and I lifted N222SK off the pavement and around the pattern. The feeling I had during those few moments were completely void of any fear—it was the most fun I had ever had. The plane flew as if it had been flying for years. Absolutely no problems whatsoever. What surprised me even more was the landing. I had never flown an airplane with a stick, much less a side stick. That first landing was the smoothest landing I've ever made. Every other landing since then has also been smooth, too. N222K cruises at a true airspeed of 180 mph at 6,000 feet and 75 percent throttle. The engine is a Continental C-90-12F turning a Bruce Tifft prop. Empty weight is 597 pounds. I am extremely pleased with the airplane. I have never flown a plane even similar to it. Like I've said many times before, it flies like an airplane should."

Non-intentioned "first flights" are also duly reported by the builders. Lee Roan of Temple City, California (Fig. 8-3), really hadn't planned to make his first flight, at least not that day, as he had not flown in the past six months and had not yet had his planned Yankee checkout. He said, "I made a high-speed taxi, got to 70 mph with nose wheel off, then was flying. I pulled the throttle back but floated past the half-way mark, still floating. I decided it was time

Fig. 8-3. Lee Roan, shown here with his fully IFR-equipped VariEze reported in the newsletter that his first flight was purely an accident. Control stick is snubbed by a bungee cord and a special Sigtronics headset is stowed between flights in the right rudder well.

to fly, so I took off and flew around the field and came in too fast the first time. So, I went around and made a good landing the next time." Roan is an electronic designer for Sigtronics, Inc., and naturally has the latest Sigtronics intercom system and headsets in his Eze, which he demonstrates at the drop of a decibel.

Worldwide interest in the VariEze shows up in the columns of *The Canard Pusher.*

From Mick Hinton, Kororo Via Coffs Harbour, Australia:

"I would like to inform you that my VariEze VH-EZH is now flying (fantastic to say the least). It weighed out just under 600 pounds with alternator and limited IFR panel.

"Thank you very much for your help, which has allowed me to build this magnificent machine. It flies hands off and I just can't stall it, but will be going back into the work shop soon to install the wing cuffs.

"After the necessary flight tests in March of this year, I was able then, accompanied by my wife, to do my longest flight in EZH and enter my Eze in the SAAA convention of the year at Bowral, N.S.W. I am very proud to say it won the Reserve Grand Champion homebuilt of Australia. This was my first attempt at building so mere words will never express how I felt that day.

"Since the air show, my phone has rung constantly with enquiries from future builders of the VariEze.

"Hope to see you at Oshkosh this year."

And from Switzerland, we hear from Dana and Rudi Kurth:

"Our VariEze flew for the first time at Sion Airfield in Switzerland. We used Sion for the tests because it has a longer runway than our local airfield at Grenchen.

"I have been pretty busy. I translated the Eze plans into German and I also translated *The Canard Pusher* for a group of Eze builders here. So what with all that, plus teaching Grenchen Tower personnel English, teaching at the local school, and a 10-month-old baby to chase after, you can see why we haven't written sooner.

"Our registration number is HB-YBG. HB is Switzerland, Y is the prefix for homebuilts here and BG is just the letters which came next in sequence—you can't choose special letters here, like they can in the U. S. So we just tell people that it means, "*H*ey *B*urt, *Y*ou're *B*loody *G*reat!"

"For your information, over forty Ezes are being built in France, ten in Italy, about five in England, about eight to nine in Switzerland, and I believe one in Holland. Further, four or five are building in Germany."

Since it was flown for the first time, we have had 78 different passengers in it, have landed at 29 different airfields, and she has been flown by six different pilots, including a Swiss Air Force pilot, the Swiss Ministry Chief Flying Instructor, and Inspector for homebuilts from the Swiss Air Ministry, the Tower Controller from Grenchen, and French Army test pilot who test-flew the Mirages. The Inspector did a belly landing (nose gear up) at Grenchen, and got out of the Eze very red-faced, but having done very little damage, to our relief, and to the amusement of the many on-lookers.

"We will be flying to Germany, France, England, Holland, Belgium and Denmark this year—a flight we had planned to Israel, via Italy, Greece and Cyprus has had to be cancelled because they have no av/gas, and we can't fly through Turkey, Lebanon and Syria. In August, we will be flying with our daughter with us. She will be 16 months old then, so we hope she's not too heavy.

"It's wonderful to be able to get into the plane and fly to Grenoble for a coffee, or to be able to fly off and land in England less than 2 ½ hours later—even SwissAir can't beat that, when you count the check-in time, luggage-finding, queue for the Customs, etc."

[A full account of the test-flying of HB-YBG is carried at the end of Chapter 4].

Much of the newsletter material is technical with detailed drawings of minute parts. Once in a great while, there's a general reminder for quality control, like this statement by Burt.

"We encourage builders to help police each other by being honest with each other and letting the guy doing marginal or submarginal (junk) work know about it. It appears that most builders are doing a good job of building their airplanes, but there are a few of you who aren't. We have seen a few examples of workmanship so poor that the parts are structurally unsound and cosmetically wretched. The unfortunate part of it all is that some of these builders either don't know or refuse to admit that they have created junk, not airplanes. It may be a bitter pill to swallow, but some of you will not be capable of building a VariEze. We have found a few builders who try and stretch the limits of our acceptability criteria so they can squeek by with sub-marginal workmanship. If you know of someone who is trying to 'slide by' with a junk airplane, it is in your own best interest to let him know about it in no uncertain terms. A poor safety record hurts us all. All airplanes are infernally complicated and difficult to build, some are just worse than others. The VariEze is 'less worse' than most, but some builders will not be capable of mastering it. If you are in this group, be honest with yourself and admit it before your pride gets you killed and puts restrictions on the rest of us. Get some help and get your workmanship up-to-speed before you continue."

In 25 issues of *The Canard Pusher* and its predecessor, there have been some gems of information and interesting questions and answers. Some examples follow.

"Use a plastic garbage can liner to keep epoxy off clothes. Cut three holes, one for your neck and two for arms." - Phil Supan.

"Be careful when storing foam blocks; not only can sunlight ruin them, but rats and mice love to dig tunnels in them!"

"Q. My wife is 6'5" tall. Can she fit the back seat?"

"A. The front seat allows 'stretch out' comfort (feet in front of the rudder pedals if you desire) for pilots to 6'7" and 210 lbs. Back seat is comfortable for pilots' passengers up to 6'5" and 220 lbs. In fact, those of you who were at the Watsonville Fly-in may have seen a 6'9", 210-lb man get in the back seat with the two full suitcases. His comment was, 'relatively comfortable.' Even he was not pressed up against the canopy."

"Q. I want my Eze to look more like a Defiant. Can I eliminate the lower winglet?"

"A. Performance-wise, yes; it only gives about 1 percent induced drag reduction. But do not leave it off— it protects the rudder and cable in case you drag a wingtip on takeoff or landing."

"Wilma Melville—possibly the only woman VariEze pilot at

Fig. 8-4. The VW prototype N7EZ went through a ditch at 75 mph—when the pilot undershot. The landing gear tore free of the airplane, as it was designed to do, since it is the easiest and cheapest part of the airplane to replace. Dick Rutan later rebuilt this airframe (courtesy Don Dwiggins).

this time—has completed her 100th hour in the airplane which she and her husband John built over a period of 1½ years. A physical education teacher from Torrance, California, she included the EAA Oshkosh Fly-in in her schedule. John built the Eze with Wilma's help, but he is not a pilot and is thus confined to the back seat."

"The 'Real' George Scott suggests a neat way to improve your proficiency before flying your VariEze. George used a Cessna 172, sat in the right seat, reclined the seat to simulate the seating position of the Eze. Then go out and practice some high-speed taxi, runway flights and landings. Do take a safety pilot with you in the left seat."

"Q. Are dual controls planned?"

"A. No. Again, I do not plan to compromise the design simplicity to do a mission other than that of efficient cruise. Learn to fly in an airplane which was designed as a trainer. Dual controls would triple the number of parts in the control system and eliminate one suitcase. Also, controls could be jammed when flying solo with baggage in the back seat. Currently there is nothing in the back seat which moves. Four pilots have been successfully checked out in N7EZ to date—none of them had side-stick experience."

"Q. What is the minimum size door in the shop to hatch a finished VariEze?"

"A. If you leave the main gear off until after the airplane gets out of your shop, a 30″ × 68″ door or window is enough."

"A few builders have asked what the weakest part of the landing gear is, and the answer is the attachment pads. Don't misunderstand, it is strong enough to do the job and then some, but it is the part we expect to break first if the system is overloaded.

On N7EZ (VW prototype), its trip through a ditch at 75 knots last summer tore the gear free of the airplane (Fig. 8-4) by failing the attachment layup, but the gear strut itself was not damaged and will be reinstalled on the airplane. We specifically tailored the attachment layup to fail first in a crunch because it is the easiest and cheapest part to replace."

"A border patrol Cessna Citation picked up a suspicious target cruising toward Florida from an island in the Caribbean. Its strange configuration was a complete puzzle to the radar operator, so the aircraft was tracked to its destination in Florida. Law enforcement officers were waiting to nab the pilot and his load of marijuana. The airplane? A VariEze."

VARIEZE NEWS EXCHANGE

A regional newsletter, the *VariEze News Exchange* (V_EN_E), has been developed by George Scott, Jr., of Cummin, Georgia. Calling himself "The Real George Scott," he is one of the charter members of EAA Chapter 611 in Georgia. We have excerpted some of the newsy items relayed by Scott.

"Well, here it is mid month again and time for another VariEze newsletter. How time flies when you're having fun.

"Yours truly, with N240EZ, spent the last two weeks of July on a tour of the midwest and on to Oshkosh with no problems

Fig. 8-5. VariEze builders check what other builders have done. Here a couple peer through the canopy of a parked VariEze to see what ideas might be used in their own ship. Note solar panel in front of canopy to power radios in daylight.

except for marginal VFR weather. It's amazing how one can fly 3,000 miles with a compass and a sectional map and not get lost (at least for very long).

"Oshkosh was a ball! Weather was reasonably good and at last count, 35 VariEzes, one VariViggen, and three Quickies were in attendance. We had one 'Flyby' of 10 VariEzes, the Long EZ, and the VariViggin, and in a tight pattern with slower aircraft, it do make life exciting!

"Interesting problem flying a VariEze cross country—I was eastbound level at 5500'. Another aircraft appears to the right also eastbound at the same flight level on a path that will cross slightly behind us. He thinks we are going backwards, and since the aircraft to the right has the right-of-way, he turns to the right to pass behind us! I'd sure like to hear his version of what he thought he saw and the evasive maneuver that followed!

"The welcome you get when flying into most any airport in a VariEze is quite interesting. They all want you back because of the interest it creates (Fig. 8-5). I've been invited into Dobbins Air Force Base for a static display and the Thunderbirds will be performing. I am also invited to Dalton, Georgia, for static display and fly by for the Lions fund raising airshow. We also plan to fly to Tullahoma, and I'll be working some on the flight line as a flagman.

"We'll get together at Tullahoma to put the final plans together for the Gainesville VariEze builders conference. I've had word from many builders and potential builders that they plan to walk, drive and fly to Gainesville to be able to talk to you experienced builders and flyers. Since the 'bull session' at Oshkosh was held near the flight line and the loud speakers were going all the time, many builders said they couldn't hear. Well, come on to Gainesville, and you'll be able to get your questions answered on a personal basis from all the VariEze builders (Fig. 8-6). But please don't ask where the tail wheel is."

"Despite the weather with fog, a ceiling of 200 feet, and less than a mile visibility until noon, we had a wonderful VariEze Fly-in at Gainesville, Georgia.

"Builders from North Carolina, Georgia and Tennessee were here and they didn't waste any time in getting specific questions answered from other builders or those who flew in. Many builders were taken on rides and even some potential builders were able to get in some rear seat stick time.

"As each new VariEze arrived in the pattern, a hush would come over the crowd and all eyes scanned skyward to see who

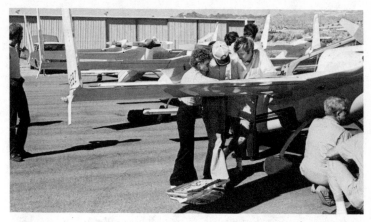

Fig. 8-6. Builders congregate and compare notes. Bruce Evans of Carlsbad, California, hosts other interested pilots during a group fly-in.

would first recognize the newcomer. Then, what a pleasure to welcome those proud pilots and crew.

"We did some formation flying on Saturday but perhaps the most interesting feature of the whole fly-in was the spot landing contest on Sunday. Would you believe that a VariEze could be landed consistently very close to a designated spot on the runway? Well, I didn't believe it either until I tried it and also saw the results of the others. Five VariEzes participated in the contest with the following rules: 1) normal pattern altitude of 800' on downwind 2) no use of the speed brake 3) add no power after turning final 4) closest distance out of two tries 5) breaking landing gear disqualifies contestant! Jeff Rose of Chattanooga won and his second try was right on the money. The last place pilot was only 20 feet away from the spot.

"I realized that I was not picking out that theoretical 'spot' on the runway and then using that spot throughout downwind, base leg and final and on touchdown. Try it and you'll make landings look 'VariEze!"

"If you don't have time for a 'Quickie,' then try it 'VariEze!' "

In his Christmas newsletter, George Scott commented:

"The VariEze rumor mill has a new model being developed by RAF! It is to be tested December 25 by an elderly gentleman with a white beard in a red jump suit. It is reported that the new aircraft was built in six weeks by 365 elves, none of which were allergic to epoxy. Stellar instrumentation negated any use for flaky gauges and the tach was reported to be accurate to ±.000994 percent.

Cruise speed is reported to be 186,000 mps on eight horsepower and fuel consumption is four bhph (bales of hay per hour).

"The plane has room for 10 lbs. of luggage in a 5-lb. sack and the front cockpit is extra comfortable since the seat is constructed of a filler of Preparation H covered by Non-itching fiberglass cloth.

"Approaches to landings are accomplished via a remote TV camera located on the non-steerable, non-retractable teflon skid. The lower .9 of the TV screen is blocked from view to simulate the visual approach of older style canard type aircraft.

"So, if a futuristic appearing aircraft makes a low pass over your house on Christmas Eve, don't call the FAA, just smile and wave.

" 'Northpole Unicom. This is Experimental 1979 EZ entering a left downwind for Runway 36. This will be a full stop and the last flight for this year.' "

"Merry Christmas & Happy New Year!

"The 'Real' Santa Claus"

TEST-FLYING TIPS

The "Real" George Scott added a collection of building and flying tips for his newsletter subscribers. He detailed prerequisites and actual test flying of his VariEze this way.

Stop! After all those hard-earned dollars and many, many hours of labor on your VariEze, stop for just a minute and reflect on your flight test procedure. Ask yourself a few questions, such as:

☐ Am I current with recent hours in a couple different airplanes such as a Yankee or Grumman American trainer and a Piper or Cessna?

☐ Do I have an up-to-date BFR and medical?

☐ Has my local EAA Designee inspected the aircraft?

☐ Have other VariEze builders inspected the aircraft?

☐ FAA inspection and airworthiness certificate?

☐ Am I following Rutan's procedures in the owners manual with updates from the 'Canard Pusher?'

☐ Is the weather perfect?

☐ Am I psychologically ready?

A technique that worked well for me was to fly the wife's 'Skyhawk' from the right seat with the seat tilted back to simulate the reclining position of the VariEze seat along with right hand on the stick and left throttle. This forces you to glance out the canopy to the edge of the runway as the nose comes up in the flare just before touchdown. Don't move your head and take a chance of causing vertigo, but just move those eyeballs and look as far forward on the runway as the instrument panel and canopy allow.

On my first flight, I requested our EAA chapter designee to fly along in his PL-I for observation. Don't try any formation flights on that first test flight, but it is comforting to have someone else up there with you and the good Lord. If you do need to make an emergency landing, the chase plane can do some radio work to clear other air traffic for you. (Remember, you are still over the airport, so a dead stick landing is 'VariEze.')

A few months ago I had a checkout ride in the rear seat of a VariEze which ended in a main gear failure. This was at a busy fly-in with about six aircraft in the pattern, and I can still remember that courteous Pitts pilot behind us as he radioed others on the frequency—'Gainesville traffic, we have a VariEze down on Runway 22; go around, go around; the runway is closed.' Even though the airframe suffered damage to winglets, prop, belly and air scoop, the craft was back in the air in two weeks!

Those of you that are close to the test-flying stage should arrange two things. First, ask for some rear seat stick time in another VariEze to get used to the sensitivity of the controls. Second, ask another VariEze pilot to check the flight characteristics of your new airframe. This is no doubt one of the most difficult things you'll contemplate and I can argue either side, but look at it this way. That pilot who already has several hours in a VariEze will know instantly as he lifts off whether or not you have an out-of-trim airplane or possibly something else wrong, where on your first flight you might over-react to a normal rotation or out-of-trim condition. What's that you say? 'I built that aircraft with my own hands and by golly I'm going to test fly it myself!' Well, okay, I can understand your feelings on that, but at least coax your friend to make some high speed taxi runs up through about 80 mph to test flight controls, trim settings, engine operation, and the like.

After your first takeoff, establish a cruise climb and, staying close to the airport, make some gradual turns to gain altitude to about 6000 feet. Then try a 500-foot per minute descent to 5000 feet for a simulated landing. Try two or three of these before your first actual landing and get used to the altitude of the aircraft at touchdown. Of course, you can't simulate the depth perception of the runway at five feet while you are at 5000 feet, but the practice will help.

Make your approach at 95-100 mph to at least a 3500' strip. This airspeed is high but will make you feel more comfortable with the visibility over the nose. A slightly circular approach from downwind to final provides excellent visibility, but don't practice this on your first test flight if you haven't made a practice of it.

The thrill of your first flight in your VariEze simply can't be put into words by this writer, but I can assure you that the euphoria lives and flies on and on!!

I'd sure like to tell you that test flying is difficult; but really it's VariEze.

TYPICAL MUTUAL AID SOCIETY

The assembly of your very own homebuilt airplane becomes a very personal thing. When you begin a project that can take from one to five years (or even longer),you're committing a substantial part of your time and complete interest. Anywhere that two or more Eze builders live within commuting distance, there becomes instant comraderie; and as the numbers grow, self-help groups develop. The San Diego VariEze Squadron is typical of these grass-roots groups. Ray Ganzer, who handles publicity for the group, provided the following information on this local "mutual aid society."

"Just after Rutan announced that he would sell plans for a Continental-powered VariEze in 1976, several potential builders met to discuss strategy—how to get plans, search for materials, etc. Al Coha, Squadron Director, has provided the cohesive force that is so necessary to keep a group of people together.

"Shortly after the first meeting, several group purchases were made, including plans, foam, landing gear and hardware. Subsequent progress on individual projects varied widely; some moved ahead rapidly, others slowly.

"Regular, but informal, group meetings evolved, usually held after each quarterly issue of Rutan's newsletter, *The Canard Pusher*. We eagerly awaited any helpful building hints, plans changes, etc., and new of completed projects and flight testing. These meetings were held in the builders' homes, with a different builder hosting each meeting.

"The completed aircraft (11 to date) are all different, not as basic aircraft, but for such things as engines, props, instruments, fairings, spinners, cuffs, lights, etc.

"Recently meetings have been scheduled once each month. With so many aircraft completed, emphasis of the meetings has shifted (from the early period when everyone was constructing) to now include discussions of flying activities and aircraft and engine maintenance. Not only are such meetings being held for technical discussions, but they are provided an excellent vehicle for flight safety education. Many wives are also attending, thus adding to the social flavor of the meetings.

"Recent activities included a trip to Mexico by four aircraft, an unscheduled arrival of five aircraft at Borrego, and the Bullhead City fly-in."

INTERNATIONAL VARIEZE HOSPITALITY CLUB

In a taped discussion between Dr. Donald Shupe and himself about the changes in their lives with the building of their respective VariEzes, Ed Hamlin commented, "The Eze has affected our friendship habits; it has expanded the horizon in friendships that we're able to establish."

"We had a group of people in our area around Auburn, California, that we associated with—airplane builders mostly. But the Eze has opened up a new facet of friendship patterns in that we've become friends with Eze builders that live very far away whom we would never have met had we not started flying the airplane."

"That was exactly the reason my wife Bernadette and I got the idea for the International Hospitality Club," explained Dr. Don Shupe, of LaVerne, California (Fig. 8-7). "We became so interested and impressed with the possibilities of travel and meeting new people. Having spent many weekends with the Hamlins, we decided that was the kind of experience more people needed to have available. So we decided that the thing to do would be to

Fig. 8-7. Don and Bernadette Shupe in their VariEze "Puff"—the Magic Dragon—during a fly-in at Mojave. Other VariEze owners ease the nose to the ground after the Shupe has retracted the nose gear.

239

Fig. 8-8. Popular defiant is in the center of the group.

establish a club where people who thought they wanted to visit other people could have a listing of other Eze builders and pilots who would be interested in having houseguests for a day or two. And so this club has become quite successful because apparently there are sufficient numbers of people who feel the same way that it made the operation viable."

The following announcement regarding the formation of the International VariEze Hospitality Club was carried in *The Canard Pusher:*

"The VariEze Hospitality Club is a non-profit organization to put VariEze pilots in touch with other pilots or builders who would like to share their homes for one or two nights with travelling VariEzers. The idea is to increase communciation and friendship between VariEzers and to provide an alternative to expensive motels and car rentals. In addition, we will hopefully have people who could help with parts and repair when needed within a hundred miles or so of every airport.

"If interested, please send a check of $2 with a clearly printed 3 × 5 card to Dr. Donald Shupe, 2531 College Lane, LaVerne, California 91750, phone (714) 593-1197.

"The card should contain your name; address; telephone number; aircraft number; hours on it or percent complete; number of people you are willing and able to accommodate at one time; other conditions, such as airport hangared, other close airports, condition and length of runway.

"People sending names must understand that we cannot guarantee that other respondents will accept you as overnight guests, but our experiences (76WN, 777EJ, 39EZ) have been very rewarding. Those who send names will receive all information on all respondents."

At this time, the club has grown to 100 members. Excerpts from roster demonstrates the international flavor and hospitality offered:

From Canada: "One extra bedroom and car hangar—beautiful country!"

From Australia: "With our 'banana plantation' we can accommodate up to four people. Come see Australia and have a banana!"

From Southern California: "2 people with car, 3 with motorcycle, 4 with bicycle, 5 with roller skates, 6 or more?"

From Northern California: "1 rollaway, 1 double bed, 2 bedrolls on floor, and if you like, bring your own rolls, too."

From Illinois: "1 or 2 people, call ahead. Airport: 100 oct., tie down, hangar space—sometimes; could usually provide transport or help repair."

From New Mexico: "2 people at a time; 3 runways over 5000 feet, hangars available; we are always eager to talk airplane or play airport!"

From Texas: "No Eze yet, but can accommodate up to 6 people; if more please call ahead. Come and try our 9100 by 150 foot runway."

Fig. 8-9. Completing a VariEze and flying it starts a whole new way of life. Visitors at Mojave compare notes and building techniques of their "proud birds."

Fig. 8-10. Twenty-five canards—count them—assemble at Mojave for a surprise birthday party for the Rutan brothers and their grandfather. This is the largest single gathering of Rutan designs yet on a regional basis, though crowds have been larger at Oshkosh.

From South Carolina: "2 people, possibly more if notice earlier; 2 airports with hard surfaces and services; 1 airport has 'grass surface' but very good condition. We are very glad to locate a UFO in S. C. Welcome!"

From Alaska: "Myself and other EAAers will accommodate any number of people! You all come on up and do some fishing. Survival gear, please!"

FLY-INS FOR CANARDS

By no means is all the designer-builder interface on paper. What is becoming an annual meeting keyed on Burt's birthday is a Mojave fly-in (Fig. 8-8). First there were only a handful of Vari-Ezes. Ed Hamlin remembers, "My first really long cross-country was an interesting odyssey because we went to California City where we met three other Ezes and put together a formation flight to Mojave in honor of Burt's birthday. At that time we had the largest congregation of VariEzes ever put together (Fig. 8-9)." [Historian's records indicate six VariEzes at the surprise Party.]

The number grew to 25 aircraft (Fig. 8-10, 8-11) that showed up for Burt's 37th birthday, the largest of the regional meetings to date (Fig. 8-12). The surprise party (Burt never remembers his birthdate) was put together by his parents, Dick's wife Emily, Pat Storch and Mike and Sally Melvill to honor three birthdays in the

Rutan family: Burt, his brother Dick, and their 91-year-old grand-father, Jessie Clyde Goforth from Fresno, California (Fig. 8-13). The cake's inscription was "Happy Birthday, Orville, Wilber and Grandpa."

VariEzes on the flight line came from near and far (Fig. 8-14). Norman Ross of Victoria, B.C., Canada, brought his grand champion Eze. Gary Johnson came from El Paso, Texas, Ray Dullen was from Tillamok, Oregon, while Charlie Richey flew in from Las Cruces, New Mexico. When you have an Eze, distances really don't seem to matter that much.

A catered lunch was served in the hangar and a magician performed "flight of hand." Builders compared instrument panels (Fig. 8-15), special modifications (Fig. 8-16) and generally swapped flight experiences. Five San Diego-based Ezes made a formation fly-by. Later in the day, Burt flew passengers in the Defiant, Dick was in the Long-EZ, Mike Melvill performed with his VariViggen, and several pilots joined up with VariEzes (Fig. 8-17, 8-18). This was strictly a homegrown fly-for-fun gathering.

Not long ago, 17 VariEzes, one Long-EZ (Fig. 8-19), a VariViggen and a couple of store-built aircraft attended a fly-in at Bullhead, Arizona, organized by the International VariEze Hospitality Club (Fig. 8-20). Members Bill and Julie Lermer of Spring Valley, California, picked the site for this fly-in (Fig. 8-21). The Ezes came from California, Arizona, Utah and New Mexico, logging a total of 9,000 miles to attend this social and mutual admiration get-together.

Jaded casino gamblers spent considerable time watching the antics of "those funny looking little airplanes" as builders swapped rides, took pictures and watched Dick Rutan perfect his aerobatic routine in the Long-EZ. Non-flying airport visitors asked all the usual questions: What are they? How much do they cost? Where

Fig. 8-11. Homebuilts by the dozens line up at Mojave, California. Virtually all homebuilders are members of the EAA.

Fig. 8-12. Visitors crowd the ramp at Mojave on an open house. Many proud builders visit Rutan's office to show him their new aircraft.

Fig. 8-13. Family portrait on a birthday at Mojave. Standing, left to right are Dick; parents, Dr. George and Irene; sister, Nellie; Burt. Seated is the Rutans' 91-year-old grandfather, Jessie Clyde Goforth.

Fig. 8-14. Eze-come, Eze-go, with this line of visiting canards parked in front of the RAF at Mojave. Workmanship and finish on most of these composite aircraft is superb.

can you buy one? One casino van driver was reluctant to drive behind a VariEze that was warming up because he didn't know which way it was going to go.

The VariEzes proved again that they require a minimum amount of parking space as 12 grouped nose-to-tail in a ramp area usually reserved for a single Beech QueenAir (Fig. 8-22).

Fig. 8-15. USAF pilot Ken Swain, Nut Tree, California, shows off his simple VFR instrument panel and side-stick installation. Captain Swain, who flies military for a living, has limited his "for-fun" Eze to a basic panel. He logged 400 hours on this ship in its first two years, including one trip to Oshkosh with his wife when she was five months pregnant—with twins yet!

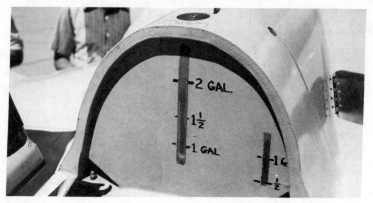

Fig. 8-16. Here's the gauge reading for the two-gallon header tank above and aft of the rear seat on a VariEze.

On both days of the fly-in, Dick Rutan worked in the new Long-EZ on his aerobatic demonstration routine. The program proved so popular during the recent Lakeland, Florida "Fun & Sun" fly-in, that Rutan has been asked to present it at the EAA Oshkosh Fly-In. At the time of the Arizona get-together, designer Burt was enroute from an Australian visit.

The International VariEze Hospitality Club (Fig. 8-23) presented awards to Jack and Marilyn Day for the longest distance flown—some 550 nm. Recognition for the most hours logged in a VariEze went to Ed and Joanne Hamlin of Rocklin, California, with

Fig. 8-17. After his birthday party, Burt Rutan flies in the Defiant over the visitors with the Long-EZ and a VariEze off each wing.

Fig. 8-18. Three of Burt Rutan's creations fly in formation near his Mojave, California factory.

more than 500 hours. An all-in-fun Pterodactyl Award went to Mike Melvill (Fig. 8-24) who flew in with his VariViggen. The Grand Champion Award for this regional fly-in went to Gary and Betsy Hertzler of Tempe, Arizona. The Galoshes Award for courageous aviating went to Charles and Joan Richey who took off in the snow from Las Cruces, New Mexico, chased an errant compass that was 30° off, and viewed some new scenery enroute to Bullhead.

Some 30 VariEzes were expected in Bullhead, but strong Santa Ana winds over the Sierra Madre Mountains discouraged a number of pilots. Dr. Shupe reported his outbound flight to be the roughest he has yet encountered in more than 450 hours in his VariEze, "Puff," decorated with a dragon design. "I was too scared to turn around," he admitted later.

Bullhead City is becoming an increasingly popular fly-in destination since a new ferryboat landing was completed directly

Fig. 8-19. Dick Rutan flies the Long-EZ over the Colorado River near Bullhead, Arizona, during a group fly-in where everybody gets into the act.

Fig. 8-20. A covey of VariEzes park at the Bullhead City, Arizona, Airport during a fly-in. Just across the Colorado River is the Riverside Motel and Casino where some pilots stayed.

across the highway from the airport office. Free 24-hour shuttles take no more than a minute to the Riverside Motel and Casino at Laughlin on the Nevada side of the Colorado River. They replace the six-mile drive over the top of Davis Dam and back down the Nevada side. Shuttle boat service is also provided between the Riverside Casino and the River Queen Motel in "downtown" Bullhead. Additional free boats shuttle from Bullhead to the Nevada Club and other new gaming resorts on the Nevada side of the river (Fig. 8-25).

First of the new resorts and casinos, the Riverside was begun in 1966 by pilot Don Laughlin. He named the area after himself and at one time had a 2800-foot flight strip partially on his property. After adjoining government land became unavailable, the shortened flight strip was closed. Now Laughlin flies his new Robertson-conversion Cessna 310 from the Bullhead City Airport and keeps his 1977 Engstrom Shark helicopter on a 12-foot-high

Fig. 8-21. Visiting VariEzes "huddle together" at Bullhead City in the parking space normally taken by a twin Beech. Mike Melvill taxies out in the VariViggen. The Long-EZ is at the left of the gas pump.

Fig. 8-22. Twelve VariEzes park nose-to-tail in a corner of the tiedown area at the Bullhead City Airport. Long-EZ is parked in the foreground. Note shadow of camera plane.

landing pad in the casino parking lot. Laughlin spent a great deal of time admiring the covey of "strange homebuilts" and talking with the Rutans.

The 4,000-foot paved, lighted Bullhead Airport has Unicom on 122.8. The best windsock in the area, however, is the plume of smoke trailing from the tall steam-generating plant just downstream from the airport.

As the fuel crunch increases, VariEze pilots tend to smile most of the time. A survey of visiting pilots at Bullhead indicated

Fig. 8-23. VariEze Hospitality Club presents awards during a weekend fly-in. Standing at the left is Dick Rutan. Ed Hamlin of Rocklin, California, and Mike Dehate of San Diego are standing at the right. Don and Bernadette Shupe are seated at the far left.

Fig. 8-24. Mrs. Isabel Melvill and her son Mike talk with Dr. George and Irene Rutan at Bullhead City. Mrs. Rutan is Historian for the International VariEze Hospitality Club.

Fig. 8-25. Bullhead City Airport and parked VariEzes are across the Colorado River from the gaming resorts at Laughlin, Nevada. A free ferry shuttles between the airport and the Riverside Casino, whose porch is shown in the foreground.

that the average VariEze with an 0-200 Continental engine (100 hp) uses 5.8 to 6 gallons per hour, with an average speed of 165-170 mph! That's 28 ½ mpg or almost 57 seat miles per gallon.

Burt Rutan's efficient canard design becomes increasingly popular, particularly for pilots making long cross-country trips. Two builders who have VariEzes now flying are seriously considering a round-the-world, two-plane flight in 1981—and their wives plan to go along!

Perhaps it'll all be VariEze.

Chapter 9
Quickie—The Engine Came First

The amazing Quickie (Fig. 9-1) is the brainstorm of Gene Sheehan and Tom Jewett, two development engineers who searched nearly four years to find a reliable gasoline engine in the 12-25 hp range with sufficient power for an efficient, single-place sport plane.

Gene Sheehan had worked in the aerospace industry since 1964 and with homebuilt aircraft since 1973. He was involved with several prototype homebuilt projects, including the BD-4, a helicopter, a gyrocopter and a BD-5. A former University of Texas student, he is also a private pilot.

Tom Jewett (Fig. 9-2) was a flight test engineer on board the Rockwell B-1 Bomber. He has spent his entire career in flight testing new aircraft ranging from homebuilts to jets and is a graduate engineer from Ohio State University. He is an active flight instructor.

Since engine development is critical to any new airframe, it came ahead of the Quickie's configuration. Gene and Tom were purposely secretive in their engine project, not wanting to follow the path of some other developers who put plans and kits on sale before the systems had ever flown or been proven.

"This is not the policy of our little skunk works at the Mojave Airport," explained the developers of the Quickie in their initial brochure. "The development of the Quickie was one of the best kept secrets in aviation. Until its first flight late in 1977, its existence was known only to a handful of people."

The "skunk works" refers back to famed Lockheed designer Kelly Johnson's super-secret skunk works that developed the

Fig. 9-1. Quickie in flight near Mojave. Gene Sheehan is at the controls. N77Q is the original prototype and is being used as a test bed for new improvements.

P-80, the famed U-2, and the mach 3 SR-71 "Blackbird." Johnson was noted for a minimum of engineering drawing and a maximum of prototype building to cut time and cost in new aircraft development.

The Quickie story really goes back to early 1975 when Gene and Tom began looking for a small, efficient, reliable engine. The search included two-stroke and four-stroke engines used in chainsaws, garden tractors, motorcycles and automobiles. The search was frustrating because lightweight, powerful engines lacked reliability and the ones that had proven reliability were either too heavy or did not develop enough power.

SEARCH FOR AN ENGINE

The search for a suitable engine was outlined by the two engineers in their company brochure.

"The origin of the Quickie began with the search for an engine that required over two years. Until it was completed, no serious thought was given as to what the aircraft should look like because the aircraft was to be designed around the engine (Fig. 9-3).

"The requirements for the engine were simple enough:

☐ 12 hp to 25 hp ☐ Low fuel consumption
☐ Lightweight   ☐ Reliable, Reliable, Reliable
☐ Small size

"Many different types of engines were evaluated prior to making that selection.

"**Two Stroke**. These engines have several desirable features including high power, light weight, and few moving parts. The disadvantages include poor fuel economy, high rpm, high vibration level, poor mixture deviation tolerance, and questionable

252

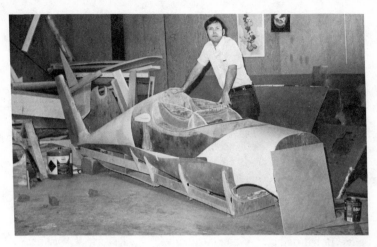

Fig. 9-2. Tom Jewett stands beside the original mold used to form the Quickie fuselage. The cabin is large enough for a 6'6", 220-lb pilot.

reliability for an aircraft application. Several small aircraft are using the McCulloch chain saw engine. It is interesting to note that all of these airplanes are either powered hang gliders or powered sailplanes, and not intended for cross-country use. Two-strokes are very mixture conscious; throttle back with the mixture leaned, descend and forget to richen the mixture, and as soon as power is added the engine is likely to seize. Failing to lean the mixture at altitude, however, may lead to plug fouling. Most dirt bikes powered by two-stroke engines have two spark plugs for each cylinder so that the rider can switch plug wires when the first one fouls.

"Rotaries. The small Sach's wankel rotary engine has many of the desirable features of a two-stroke, and it is certainly smooth running. However, these engines have had seal problems when run for long periods at high power settings, and the fuel consumption characteristics are poor in the rpm range necessary for good propeller efficiency. Besides, the engine is no longer produced.

"Four-Stroke. These engines are the best ones for aircraft use. They have a good fuel economy and tend to be very reliable. In the low horsepower examples, however, they tend to be heavy, or to require a high rpm, in order to produce sufficient power. One of the four-stroke engines that Quickie Enterprises tested was a Honda CB-175 motorcycle engine. Initially, it was too heavy, but after removing the transmission with a bandsaw and deleting all other non-essential parts, the weight was reduced to about 65 lbs.

This engine produced about 18 hp at near 9,000 rpm. While Honda engines have a reputation for being very reliable, the drastic surgery required to reduce size and weight could very well have weakened the crankcase and, therefore, reduced the reliability.

"One might ask at this point why not use a reduction drive system with a light weight, high rpm, four-stroke or two-stroke engine? There are several reasons not to do this, including complexity, cost, and torsional vibration. Given enough time, money, talent, and luck, these problems can be overcome. Often, however, the solutions only complicate the aircraft further. For example, a clutch is often used to solve the torsional resonance problem, but then the engine must use an electric starter, which adds about 25 lbs of weight.

"Volkswagen Engines. A number of homebuilt aircraft have flown using VW engines. However, a stock VW typically requires considerably more maintenance than a normal aircraft engine. This is probably because few automobile or motorcycle engines are designed for the type of continuous, high speed operation necessary for an aircraft.

"Industrial Engines. These engines tend to be very reliable, but also heavy. Most are designed to run near rated power for extended periods and usually are so dependable that oil temperature and oil pressure gauges are omitted. They have reasonable fuel consumption and frequently operate under extremely harsh conditions. Until recently, they were prohibitively heavy, and the single cylinder models have excessive vibration for an aircraft.

Fig. 9-3. Tom Jewett checks the instruments on the Quickie as its Onan engine warms up on the ramp at Mojave.

Fig. 9-4. Onan engine on a bench at the Quicke factory awaiting modifications.

"The engine selected is a four-stroke, horizontally opposed, two-cylinder, direct drive type used in various industrial applications at a continuous 3600 rpm (Fig. 9-4).

"The Onan Company has made over 1,000,000 two cylinder, horizontally-opposed, four-stroke direct engines in the last thirty years for applications from electric generator sets to snow plows. They recently introduced some aluminum versions of their cast iron series of engines. These aluminum engines weigh 98-106 lbs in the stock configurations, some 50 lbs lighter than their cast iron counterparts.

"After careful examination, it was determined that we would reduce the weight to slightly more than 70 lbs dry (Fig. 9-5). While this may seem excessive for the produced 18 hp, they are very well built. Further, if the aircraft is carefully designed around the engine as was the Quickie, the results are most satisfying (Fig. 9-6).

"Some design features are as follows:

H.P.	18 @ 3600 rpm
Type	2-cylinder, horiz.-opposed, four-stroke
Bore	3.250"
Stroke	2.875"
Displacement	47.7 in^3
Compression	6.6:1

"The manufacturer recommends up to 1000 hours between major overhauls for a normal industrial application. At this time, there is not enough data to state what the TBO in an aircraft application for a Quickie engine will be. However, it should be noted that in comparison with most industrial applications, the aircraft environment is cleaner and owner maintenance more regular.

"Much testing has been accomplished in the areas of induction, exhaust, cooling, mounting, ignition system, and the engine-airframe compatibility. The result of all this testing is an engine specifically intended for installation in the Quickie (Fig. 9-7). It is definitely not the same engine one can buy from the local Onan dealer (Fig. 9-8)."

Only after the basic engine research and testing was well in hand did Gene Sheehan and Tom Jewett approach Burt Rutan, an old friend, to develop an airframe tailored around the Onan engine. Rutan was impressed by the demonstrated reliability of the engine and began putting lines on drafting paper. Early attempts were unsatisfactory because a low enough drag in a conventional or VariEze configuration would require a retractable gear with its associated weight and complexity problems. Most pusher configurations had only a narrow range of pilot weights.

Fig. 9-5. Tom Jewett holds basic Quickie powerplant. Basic engine weighs about 70 pounds and produces 18 hp at 3600 rpm.

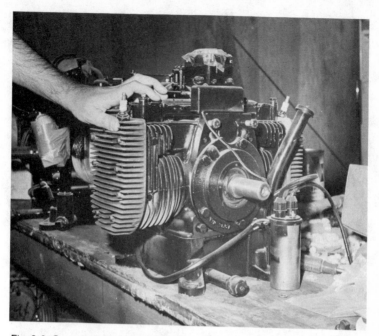

Fig. 9-6. Onan engine with Quickie modifications ready for delivery to consumers.

Rutan finally came up with a novel tractor canard tailless "biplane" configuration (Fig. 9-9). The pilot sits near the center of gravity (Fig. 9-10). The combined canard and landing gear has low drag and saves both weight and complexity. This compactness lends itself to a "glue together" airplane that saves weight on wing attachments. Full-span elevator/flaps went on the canard with inboard ailerons on the rear wing. Originally the tailwheel fairing was the only rudder (Fig. 9-11).

Once the concept was established, a detailed plane was agreed upon. Tom Jewett and Burt Rutan did the detailed design while Gene Sheehan (Fig. 9-12) continued engine development. Most of the actual construction was done by Gene who had no prior experience with composite construction. This was a simple way to prove out the concept that the Quickie could be put together successfully by a first-time builder.

The construction phase took just two months. All three developers , Burt, Tom and Gene, flew N77Q on the first day after completion (Fig. 9-13). Then followed a five-month flight test program to assure that the unusual configuration, coupled with a new-to-aviation engine, would do the job satisfactorily.

Fig. 9-7. Onan engines modified for the Quickie await delivery to homebuilders. Tom Jewett in the background looks over the assembly.

The extensive flight test program covered almost 150 hours and concentrated on the following areas, as reported in the newsletter:

Fig. 9-8. Detailed photo of Quickie engine with modifications to the industrial engine.

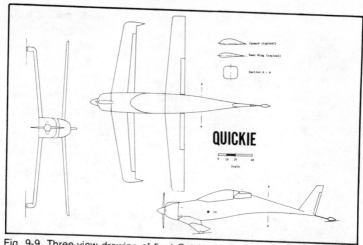

Fig. 9-9. Three-view drawing of final Quickie configuration (courtesy Quickie Aircraft Corp).

"1. Basic performance and stability and control testing throughout the entire center of gravity range at gross weights ranging up to 515 lb., which is 35 lb. over the design gross weight.

"2. Flutter testing to 162 IAS at 6,000' (about 180 TAS).

"3. Fuel economy measurements with a very accurate fuel totalizer have confirmed that at maximum cruise speed (121 mph), the fuel economy is 80 mpg. At economy cruise, the fuel economy exceeds the magic 100 mph figure.

"4. Stall/departure/spin testing; the Quickie prototype could not be made to spin during any phase of the test program.

"5. Engine and system durability and reliability; except for normal maintenance, no work has been necessary on the modified Onan powerplant. Further, no powerplant failures have occurred.

"6. Landing gear energy absorption tests to FAR Part 23 certificated aircraft standards.

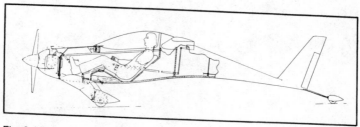

Fig. 9-10. Cutaway drawing of the Quickie sideview (courtesy Quickie Aircraft Corp.)

"7. Static load testing of the entire airframe to FAR Part 23 standards in the Utility category.

"8. Crosswind and turbulence testing; the aircraft has been operated in 50-knot winds.

"9. Independent pilot evaluations; both Peter Lert of *Air Progress* and Peter Garrison of *Flying* have evaluated the Quickie.

"The maximum speed obtained to date has been 123 TAS at 3,000'. Extrapolating this data to sea level yields a speed of about 126.5 mph. Since our prototype is "dirty" in many areas, a well-built Quickie might very well exceed these figures.

"All of our performance data is based on the average of many flights and test points.

Fig. 9-11. Final design of Quickie fin, rudder and tailwheel. Linkage from the rudder goes directly to the tailwheel.

Fig. 9-12. Gene Sheehan, who weighs 200 pounds, dwarfs the Quickie that weighs only 240 pounds empty. Onan engine idles smoothly before flight.

"It is interesting to note that an airplane that obtains 1 mph cruising speed per 1 hp (e.g., the Mooney 201 goes 195 mph on 200 hp) is considered to be a very efficient design. The Quickie obtains nearly 7 mph crusing speed per 1 hp!

"The flight test program has resulted in several configuration changes. After testing a slotted flap arrangement on the original canard, it was decided to build a new, higher aspect ratio (i.e., skinnier) canard of greater area, using a plain flap. This change set the flight test program back about 30-45 days. While the slotted flap on the original wing did lower the stall speed by about 6 mph, it resulted in rather abrupt stall charactertistics, which we felt were undesirable in light of the excellent stall characteristics that we had enjoyed with the original wing and the original plain flap. The slotted flap also reduced the basic stability margins, resulting in an aircraft that was more sensitive than the original configuration had been.

"In addition to the higher aspect ratio canard, 16″ of span has been added to the rear wing. The resulting reduction in induced drag, and increase in wing areas from these two changes, has resulted in a dramatic improvement in performance with the 18-hp engine:

"1. The current aircraft at 5,000′ MSL will climb 160′ per minute at full power at minimum speed. The corresponding value at sea level would be 300′ per minute. Maximum climb is 425 fpm.

"2. The aircraft will still climb at minimum speed at 9,000' MSL.

"3. The Quickie was still climbing at 12,000' at gross weight without a mixture control.

"4. Fuel economy of 80 mpg at 120 mph and over 100 mph at a slower speed has been verified."

At the conclusion of the flight test program, the design was "frozen" and production was begun on a homebuilder's kit. The complete kits include: 1. Production plans, owner's manual and a one-year subscription to the Quickie newsletter. (Any revisions or updated material are printed in this newsletter along with reports of other builders' experiences). 2. All raw materials for the basic aircraft, including fiberglass cloth, resins, foams, sheet metal, tubing, hardware, etc. 3. Premolded components include a clear canopy, cowling and S-glass fiberglass tailspring. 4. All welded parts. 5. VFR instruments. 6. Machine parts. 7. Tools, including mixing cups, brushes, stirring sticks, filter mask, gloves and special scissors for cutting cloth. 8. Electrical system; a 15 amp alternator, wiring, switches and fuses. 9. The 18-hp Onan engine modified for aircraft use. 10. Completed propeller.

Sales of the Quickie kit increased steadily (Fig. 9-14). Shortly after the kit program began, the Quickie received the coveted Outstanding New Design Award from the Experimental Aircraft Association (EAA) at the annual EAA Oshkosh, Wisconsin fly-in. In presenting Quickie Aircraft Corporation with the award, EAA stated that the pioneering of the Onan engine together with an exceptionally efficient aircraft design in order to bring the cost of ownership and the cost of flying down to an affordable level represented a significant breakthrough.

Fig. 9-13. Unusual configuration of the Quickie shows up in this formation photo taken near Mojave, California, with Gene Sheehan at the controls.

Fig. 9-14. The Quickies on display at the EAA Chino Fly-in. Non-builders can purchase partially completed or completed homebuilt projects from other builders. The EAA's "Sport Aviation" carries many pages of classified ads of homebuilt for sale.

FLIGHT TO OSHKOSH

The trip to Oshkosh was described in the newsletter as follows:

"The Quickie was the lightest and lowest horsepower aircraft to fly to Oshkosh in 1978. We (Sheehan and Jewett) firmly believe that any aircraft which is not flown cross country to Oshkosh should not be offered for sale to the general public as an aircraft.

"Our trip was spread over 2 ½ days, with overnight stops in Albuquerque, New Mexico and Kansas City, Missouri. The 2025 miles were covered in about 19 hours (against the proverbial headwinds!) while averaging 65.1 mpg, also a record. That means that the trip cost us about $30 in gas and one quart of oil! The takeoff from Albuquerque was made at a density altitude of 7,000'. The highest altitude reached was 13,500' west of Gallup, New Mexico. The normal cruise altitude was 7-8,000'.

"The trip was both routine and uneventful. Our biggest problem was minimizing the time spent on the ground when we stopped for gas. Usually we had to spend at least 30 minutes talking to the crowd that invariably gathered. In Dalhart, Texas, we had to wait an additional 30 minutes so that the line girl could go home and get her camera.

"For a companion aircraft, we took along a Grumman Trainer. We had originally intended to use a Cessna 150 for the flight, but found that it wouldn't keep up with the Quickie! The Grumman is about five knots faster than the Quickie and made a good companion aircraft.

"We arrived at Oshkosh two days before the fly-in started so we could relax and take a short vacation. Wishful thinking! From the time we touched down until we left, we were surrounded by people wanting to see the aircraft and ask questions.

"It was not unusual during the week at Oshkosh to find a crowd four people deep surrounding both the Quickie on the flight line and our booth in the main exhibit building (in fact, some people complained that they couldn't find our booth).

"Gene, Tom and Burt gave forums on the Quickie on both Monday and Friday. The crowd estimate on Monday was over 900 people. As a result, the forums ran long past the scheduled hour.

"We were fortunate enough to acquire a flight demonstration slot immediately prior to the airshow on several days. Quickie flight demonstrations were flown by Tom and Burt. When traffic permitted, both flew the aircraft within a box about ¾ mile long by ¼ mile wide by 500 feet high to show off the extreme maneuverability of the Quickie.

"Peter Lert (pilot report in June 1978 *Air Progress*) flew the Quickie for a photo session with *Popular Mechanics*. After returning, we asked him in front of a large crowd how he liked the aircraft. His reply was, "Flying a Quickie is the most fun a person can have in public during the daytime!"

"The trip home from Oshkosh to Mojave, California, was as uneventful as the trip East. The most important news is that we stopped at Ames, Iowa, to test the Quickie off of a grass runway. We loaded the Quickie to 20 lbs over gross weight and took off at a density altitude of about 2,000′, and a relative humidity of about 85 percent. The Quickie was off the ground within 100′ of what the Grumman Trainer required. Ames has a typical midwestern grass runway; it was rolled about two years ago, is fairly level with no large ruts, and the grass is clipped to within about three inches. Based upon our experience there, we have no hesitation in recommending the Quickie for operation off an airport of this sort.

"We also stayed over one day in Minneapolis so that some of the Onan employees would have the opportunity to see the aircraft."

In explaining the role of the Quickie in aviation today, the two developers put it this way:

"The Quickie is not intended to be an aircraft for everyone. A Quickie will never win the World's Aerobatic Championship, and it should not be outfitted with wing deicer boots and complete avionics so that it can fly IFR; nor is it the perfect airplane for the pilot that weighs 270 pounds, unless he is willing to go on a strict diet while he is building one.

"A Quickie is a fun aircraft; it is a reasonable aircraft for today; it is a creature that brings the exhilaration of flight to individuals unable to afford the machines turned out by Wichita; it is an airplane that a pilot can measure himself against—it does not fly so high that man needs help breathing; it does not require an A&P mechanic to keep it in perfect order, and it does not require a 10,000-hour pilot to utilize its maximum capabilities.

"It is sport! Some people race boats, others race cars; some spend $20,000 for a motor home, others buy snowmobiles; some buy motorcycles that are capable of speeds more than twice the national limit, others collect rare stamps—we just like to fly airplanes.

"It is inexpensive. While your earthbound friends struggle to obtain 30 mpg at 55 mph, you triple that figure at twice the speed; instead of a torturous nine-hour drive from Los Angeles to San Francisco, you take four hours enjoying the California coast.

"A Quickie is meant to go places and do things—faster than a speeding locomotive, able to leap tall mountains in a single bound, in great style, and always with much fun and amusement.

"The Quickiest way to meet people in an unfamiliar town is to fly into the local airport in a Quickie. People can't resist; they're jealous; they hold you in awe for flying this strange airplane; they know secretly that in the game of one-upsmanship, they have been topped.

"If you can accept the Quickie in this spirit, you will never be disappointed, and you will be hard-pressed to find a sport that will give you more fun for less money."

ECONOMICS OF A QUICKIE

And the developers explained the economics of owning a Quickie very explicitly.

Many pilots who may be considering purchasing or building an aircraft look only at the initial purchase price when considering how much the aircraft will "cost" them. This is a fallacy since the owner will usually spend more on maintaining a typical aircraft than he spent to obtain it in the first place.

"Most pilots will agree that it is difficult to find a production aircraft cheaper to fly than a Cessna 150. Let's compare the cost of fuel and oil for one year of a Cessna 150 and a Quickie. We will assume that each aircraft flies 200 hours a year. Since the Cessna burns 6.1 gallons per hour, as opposed to the Quickie which burns 1.5 gallons per hour, the Cessna uses $4.6 \times 200 = 920$ gallons of fuel more per year than the Quickie. At current prices, that is over

$800 more per year to operate the Cessna. In addition, the Cessna uses a quart of oil every 10 hours, whereas the Quickie uses a quart every 50 hours.

"To overhaul a Cessna 150 engine will cost about $3,000; to buy a *new* Quickie engine will cost less than $1,000.

"Since the Quickie lacks complex systems (Fig. 9-15), and since the owner of a homebuilt can legally do all of his own maintenance, a large savings is realized in maintenance cost over the Cessna 150 owner who pays about $17.00 per hour shop rate, and maybe $200 to overhaul the carburetor. We all know about the inflated prices of components with "Aircraft" stamped on them. Remember, maintenance costs are proportional to the initial purchase price, not the market value.

"A frequent comment heard at Oshkosh concerning the Quickie was, '. . . it would be perfect for me if only it was a two-place.' a two-place quickie would be considerably more expensive and time consuming to build and maintain. We think that pilots should revise their thinking; don't buy more aircraft than you need for 75 percent of your flying. If most of your flying is alone, don't buy a Cherokee Six because you take your family on one two-week vacation every year. Instead, rent that 'Six' for the two weeks and fly something much more economical for the rest of the year. You will be surprised how much money you will save. An additional benefit is that a Cherokee Six is not fun to fly, but a Quickie is! With inflation and the cost of energy skyrocketing, it's time to become more practical. If most of what you want in an aircraft is an inexpensive, safe, fun-to-fly aircraft, buy a Quickie and rent a two- or four-place aircraft when you need it."

Fig. 9-15. Cockpit installation of the Quickie. Round knob in the center is a temporary installation of a controllable pitch propeller drive. This control would be relocated in production planes since it would be a hazard in case of an off-field landing.

Yet another way to conserve flying dollars is the development of a modification to make the Quickie trailerable.

"We have completed all the necessary engineering, development and drawings on the aft fuselage cut to make the Quickie trailerable. We estimate that the materials required will cost the builder about $20.

"The technique basically involves making the aft fuselage behind the main wing removable (estimated weight 15 lbs.) so that the distance from the prop to the joint is eight feet. The aircraft can then be towed down the highway with the wings pointing in the direction of travel. Since the Quickie only weighs 250 lbs empty, a very light-weight trailer can be used, and the entire operation becomes a one-man task. Obviously, the savings in tiedown fees or hangar rents will be considerable.

"This method of making the aircraft trailerable may seem unusual, but it is much simpler, lighter and less expensive than trying to make the wings removable. Since the Quickie doesn't have a tail in the back, the loads that the joint sees are very small, making the technique much safer than making the wings removable. Nutplates and screws are used to join the fuselage parts together."

Answers to frequently asked questions were supplied by Sheehan and Jewett.

"1. *What is the TBO on the engine?* We now have over 250 hours on N77Q in addition to much more time on the test stand. We have seen nothing that would indicate the engine won't go to the 800-1,000 hours between major overhaul that other industrial users of the engine obtain in applications like electric welders, concrete coring machines, etc.

"2. *What are the "G" limits?* We call the Quickie a Utility category aircraft, which means 4.4G positive. The design limits on the wings are 12G plus; the canard was tested to 12G and the main wing to 6.8G with no sign of failure.

"3. *Is The Quickie aerobatic?* Well, it won't spin! It should have the same capability to loop and roll that a Cessna 150 does. Since we don't consider a C-150 to be really aerobatic, we don't call the Quickie aerobatic.

"4. *How big a pilot will fit in a Quickie?* Before Oshkosh, we considered 6'5" and 210 lbs to be the limit; but then we fit a 6'6" and 220 lb guy in who wanted a Quickie so bad that he was dieting (and already had lost 15 lb). He was comfortable enough that he ran right into our booth and bought an aircraft (Fig. 9-16).

"5. *Will the continental A-65 engine fit?* Only if you want your Quickie to sit on its nose in the hangar rather than fly! The A-65 is 100 lbs heavier and that is out of the question.

"8. *How about a sawed in half VW?* We talked to Van's Aircraft, which has been developing an aircraft around that engine for three years. He states that it is not an aircraft engine yet, and it will require much work, money and talent to become one.

"7. *I've never built an aircraft before; can I build a Quickie?* We firmly believe that the Quickie is the easiest-to-build real airplane in existence. There is no reason why anyone with a basic determination and willingness to learn should not be able to complete and fly his Quickie.

"8. *Is the 400 hour construction time for a professional aircraft builder?* No! It represents the average individual.

"9. *Can the Quickie be flown in the rain?* Rain has an effect on any aircraft, and the Quickie is no exception. On a Grumman Tiger, for example, trim the aircraft for cruise at 10,000 feet straight and level. Next, enter very light precipitation; the nose will drop and the descent rate will be 250 ft/min if you do not retrim. If the precipitation increases in intensity, the descent rate will increase to 500 ft/min if you do not retrim. Retrimming to level flight results in a loss of about 2 mph in airspeed.

"10. *Is the Quickie so finely balanced that grass on the canard will require almost full opposite aileron to correct?* Come on guys! That question is ridiculous, but we asked one of our builders, Delbert Whitehead, who flew his Quickie to Oshkosh, and he

Fig. 9-16. Gene Sheehan almost fills the Quickie's cockpit before flight at Mojave. Obsolete four-engine jet is parked in the background.

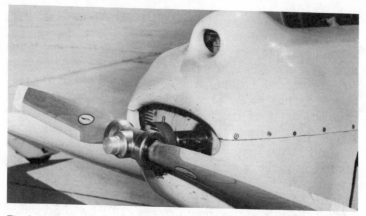

Fig. 9-17. Experimental controllable propeller fits in front of the Onan engine in the Quickie. If this propeller passes extensive flight testing, it may be offered as an option.

replied, 'I have used my Quickie as a lawnmower!' In addition, both Garry LeGare and QAC have cut grass with Quickies without problems.

"11. *Will the Quickie perform at high density altitudes?* All handbook performance data is accurate for N77Q. Our builders have, in general, seen better performance than N77Q. We have twice flown the aircraft to Oshkosh. These trips have included 7,500-foot density altitude takeoffs as well as routine cruise at 11,000'."

As more and more Quickies are completed, it is predictable that some unusual incidents would develop. For example, famed airshow pilot Bob Hoover flew a Quickie before Mojave Air Race spectators and rolled the ship in front of the crowd. Hoover described the flight as "absolutely delightful."

The Experimental Aircraft Association (EAA) asked if the developers would consider donating N77Q to the EAA for display at the museum. They indicated that they were working with the National Air and Space Museum in Washington D.C. trying to put together a display of homebuilts for the national museum, and that they would very much like 77Q for that honor.

As a result of that conversation, plus some thinking, a second Quickie is being built. As our builders have discovered, there is a significant performance improvement possible with careful attention to details (Fig. 9-17).

Another reason for building a second aircraft is that the builders are getting embarrassed by how much better the homebuilt

Quickies perform compared to 77Q (Fig. 9-18). On the other hand, 77Q is the prototype, and prototypes should never perform as well as the homebuilt versions, unless the manufacturer fudges the brochure data.

The builders plan to use 77Q for several record attempts before donating it to the EAA, and then use the second Quickie for some further attempts.

FLYING THE QUICKIE

"Flying the Quickie is the most fun you can have in the daytime in public without getting arrested," begins the flight report by Wayne Thoms publisher of *Plane & Pilot* magazine, entitled "A Plane for Under $4,000." The article is reprinted from *Mechanix Illustrated* magazine, Copyright 1979 by CBS Publications, Inc. with permission.

Among other things, the Quickie is the lowest-cost plane you can buy, coming in at under $4,000. What you get for that price in these days when even bargain-basement Cubs run to $16,000 is, of course, a kit.

But that leads to yet another nice thing about the Quickie. The kit represents the easiest-to-build plane in this country, if not the world. And it's complete even to the engine, requiring only paint and a motorcycle battery before you, too, can be up there winging with the birds.

Moreover, the Quickie is an efficient craft aloft, able to cruise at 121 mph and, when you back off to an even 100 mph, she can go 85 miles on a gallon of fuel.

We flew the only Quickie in existence at this writing—the prototype—but in shops and backyards around the country no less

Fig. 9-18. The main wheels of the Quickie are literally located in the wingtips. The unique planform of the Quickie shows up in this banked air-to-air photo. Wayne Thoms flew the Quickie over this area of the Mojave Desert.

than 70 of these little jobs are being stuck together. Many are nearing completion. Construction time is estimated at 400 hours, which is quite low for a homebuilt.

We arranged to take the Quickie up early one morning at the headquarters of the Quickie Aircraft Corp., Bldg. 68, Mojave Airport, Mojave, California 93501, which is located in the desert about 100 miles north of Los Angeles. The air would be relatively calm at that time of day. There's not much in Mojave except gas stations, a couple of motels and some fast-food stops. Just to the west lie the Tehachapi Mountains.

This paucity of population makes the town's airport a beehive of experimental aircraft activity. The flight line is filled with everything from ancient warbirds to new aircraft in semi-secret development. There's even one outfit that flies surplus jet fighters on brief missions. They test new bombs on the desert range for ordnance makers before submission to the military.

When we learned that we'd be flying behind an 18-hp, c-cyl. Onan industrial engine in an airplane that weighs no more than 480 pounds with pilot and eight gallons of fuel, we were apprehensive at first. However, one look at the plane's graceful and unusual shape helped dispel much of the uncertainty.

More significantly, the reputation of the Quickie's designers and builders—Burt Rutan, Tom Jewett and Gene Sheehan—is excellent. Jewett was a flight test engineer aboard the Rockwell B-1 bomber. Sheehan, a general aviation pilot, did much of the engine adaptation and built the first Quickie. After initial flight tests, Rutan returned to other aircraft development programs, leaving Jewett and Sheehan to market the kits.

The briefing we got at Mojave that morning was thorough. Jewett and Sheehan were anxious to insure the safety of pilot and plane. Once aloft in a single-seater, there's little margin for error.

We learned that we were to be the seventh pilot to fly the tiny aircraft, which is a taildragger. Virtually all light planes today have a steerable nose wheel and run tail-up on the ground. Going back to a taildragger was like regressing from a 747 to a DC-3, though the Quickie at least has its rudder pedals linked to the tailwheel for control during taxiing.

The main wheels literally are in the wingtips. You can't tip the plane over on the ground. In fact, an emergency stopping maneuver is to jam hard rudder and groundloop. It's a bit hard on the tires, but stops the plane.

271

We learned that the 47.4-cu. in. four-cycle engine can operate continuously at 3600 rpm, which means that the 42-inch propeller is direct-drive. There's a 15-amp alternator for the electrical system but starting is done by hand-propping.

We swung open the bubble canopy and stepped over a 34-in.-high side—about motorcycle-seat height. Seating reminded us of a small formula race car, snug and almost supine. There was good back support and a comfortable headrest. Controls are in the armrests of the cockpit sill area—throttle on the left and fighter-style side stick on the right. It makes for relaxed flying, but full deflection on a side stick normally is so slight and response of the Quickie was so . . . well, quick, that we found it easy to overcontrol.

The Quickie's design is unusual but has a purpose. The front, or canard, wing contains the elevators, while the rear wing has the ailerons mounted inboard. There's a small rudder on the vertical fin. The wing design means that conventional stalls are impossible. We were told it was impossible to spin the Quickie during flight tests. Both factors mean considerable safety for low-time pilots. What happens in an attempt to stall is that when the canard stalls, the nose drops so that a couple of miles per hour of speed are gained. The rear wing is never permitted to stall.

In practice this is what's called a pitch-buck oscillation. At full throttle, full aft stick, the Quickie actually climbs about 150 feet per minute in its strange oscillation. In fact, we were advised that we could take off and climb in this condition. We didn't try it.

Normal starting means tying the tail wheel, cracking the choke, switches on, and spinning the prop. After a brief engine warm-up, the pilot fastens the canopy, someone releases the tail and the plane taxies to the runway. Because the ignition is single there is little to check before takeoff except for controls free, carb heat off, fuel on, trim in neutral, canopy latched.

Line up with the runway, push full throttle, hold slightly aft stick and the Quickie levitates at 53 mph, flying off more or less level and going up like a slow elevator. Acceleration feels similar to that of a small two-place trainer. Takeoff distance is 660 feet at sea level, about normal for small aircraft and slightly less than a 100-hp Cessna 150.

Instructed to climb at 70 mph, we lowered the nose shortly after takeoff to accelerate. The idea was correct, but the Quickie is so much more responsive than the craft we usually fly that we pushed too much forward stick, then too much aft. The effect was an interesting porpoise at about 30 feet—that's right, 30 feet!—until we worked out the technique of gentle control movements.

Rate of climb is 425 fpm, but there's no gauge to indicate this. The instruments built into the canopy panel are airspeed, altimeter, compass and voltmeter. Tachometer, cylinder-head temp and oil temp and pressure are on the left. Our test plane had a ball to indicate coordinated flight but it did not agree with the seat of our pants. After a while we ignored it and later on the ground we were advised that our pants were correct.

Once the Quickie levels off, the speed builds slowly behind 100 mph indicated. The advertised cruise of 121 mph true airspeed can be achieved, but we weren't off on a cross-country trip and didn't try.

The idea of being strapped into a powered flying machine that weighs little more than the pilot and has the ability to respond instantly to the pilot's wishes was mind-bending. Never has this pilot felt more in control of his destiny than with the Quickie.

Approach to landing is made between 70 and 75 mph, adjusting throttle as required to reduce speed and maintain glide angle. Contact is made at about 55, tail wheel first with stick full aft. In theory, at least, this is simple. After all, the main wheels are clearly in view on the wingtips, and with a long runway there is no reason to drop the airplane in. A kiss-soft landing should be within the grasp of even a novice pilot.

We goofed slightly, and we were glad that the main gear/canard wing is stressed to 12 G's. Our drop was only a matter of inches but it felt like much more. The canard took up the shock, we steered carefully with rudder pedals and applied the brakes.

There were steep turns to be tried, power-on and power-off attempts to stall, then approaches to a landing with full-power low passes along the runway. The Quickie doesn't have enough power to be aerobatic, but it's an absolute delight to fly in every other mode. This, we thought, is how aviation ought to be: no radio, no electronic navigation aids, no controllers—just the pure, sweet pleasure of flight, one person perfectly in tune with his craft. Without question, we'd call the Quickie the ultimate recreational vehicle.

The construction kit for the Quickie is about as complete as anything in this field could be. The engine even includes an hour of dyno test time.

To sum up, the Quickie has achieved its design objectives: to be an easy-to-fly, easy-build, safe flying machine that is low in cost and quite possibly the ultimate adult toy."

Chapter 10
Defiant—A Promise For The Future

Within a few days of the Defiant's first flight, Rutan had a newsletter on the way to his builders. Some of his youthful enthusiasm spills over in the words describing this first test flight.

"Our new light twin made its first flight on June 30, 1978, with Burt at the controls (Fig. 10-1). Within one week Burt and Dick had logged enough flying for FAA to remove its area restriction and had obtained all basic performance verifications. The only maintenance or adjustments required has been changing the stiffness of the nose gear steering pushrod. This has been the cleanest initial test program we have seen on any type aircraft.

"Curtis Barry, Port Jervis, N.J. won the Name-the-Plane contest. He added that Defiant infers, 'the aircraft defies all the common assumptions about current production twin engine aircraft—in pilot skill required, safety, performance, construction, and handling.' We waited until after the aircraft had flown to name it, as we wanted to be sure it did indeed meet the above definition. As those of you that have recently visited know, we are extremely excited around here, since we are finding that Defiant actually is exceeding the estimated performance and is verifying the no-procedure-for-engine-failure design goal. It doesn't take a lot of study to realize the impact on flight safety of a twin that not only has no appreciable trim change at engine failure, but requires *no pilot action* when it does fail. You can fail an engine at rotation for takeoff or during a go-around in the landing flare. The pilot does nothing; he climbs out as if nothing happened. He has no prop controls to

Fig. 10-1. The Defiant makes a low pass near a crowd at the Mojave Airport. Even in hot desert air, there are no cooling problems.

identify and feather. He has no cowl flaps to open, no wing flaps to raise, no minimum control speed to monitor (he can climb better than the other light twins even if he slows to the stall speed), no retrimming is required; he can even leave the gear down with only a 50 fpm climb penalty. The only single engine procedures are the long-term ones: (1) cross feed if you want to use all fuel on operative engine, (2) magnetos off. Note that, in general, you do not lose the alternator or vacuum pump on the failed engine since the engine windmills at 1000 rpm (fixed pitch prop). Why no cowl flaps? The two updraft cooling systems (Fig. 10-2) were designed to have large positive cooling pressure increases with power and angle of attack. When you are cruising, the cylinder head temps stabilize at 370°F. If you then fail an engine and execute a full power climb and slow to best single-engine climb speed, the full-power engine will cool to 350° with no pilot action (same mixture). The engine installations are simpler than the most simple single. Baffling is less complex, blast tubes for mags or ram air plumbing for carb and carb heat are not required. There are no oil coolers. Oil temps run to 200°F during a climb to 15,000′ and stabilize at 190°F at high cruise at outside air temperatures of 40° above standard conditions!

"The most surprising good news is that the Defiant does not have the annoying, loud, out-of-sync noise common to the Skymaster. The pilot has to split the throttles considerably to detect out of sync at high power, and at low or medium power the sync noise is not detectable. The sync noise is more objectional on other twins

Fig. 10-2. Nose-to-nose with the Defiant. Little or nothing has been done to improve this design concept since it first flew in mid-1978.

than Defiant, even though they use constant speed props mounted on wings. We feel the main reason is the high damping of the Kevlar/wood props and composite structure. Using Flight Research mufflers, the Defiant makes less perceived noise for a ground observer than the average medium performance single.

"The airplane is a stable IFR platform (Fig. 10-3, 10-4), with less trim changes than conventional twins. It has a very solid 'big airplane' feel. Approach speed is 75 knots at light weight and 85 knots at gross. We are withholding detailed performance data until it is completely generalized and presented for all weights, but the following is typical of that being obtained: cruise at 65 percent power (maximum cruise) at 12,000' is 188 knots (216 mph) without wheel pants. Single engine climb gradient (foot increase per mile) is almost twice that of the new light-light twins at any given loading condition. Single engine service ceiling is well above these aircraft even with the gear down and the airspeed 15 knots off the best climb speed, and this is obtained *instantly*—not after a clean-up procedure! We gave Joe Tymczyszyn, FAA test pilot from the Los Angeles AEDO a ride that included single engine go-arounds initiated in the landing flare. His comments: 'Unbelievable, single engine procedures are refreshingly simple.'

"The Defiant is big inside (Fig. 10-5)—2 inches wider elbow room, 8 inches longer cabin, 6 inches more knee room in back seat and 3 feet 3 inches more baggage volume than the Beech Duchess. To get to gross weight in a Defiant you can top the tanks for 1100

NM range, add four 175-lb adults and add 75-lb baggage to an IFR-equipped airplane.

"We have no plans to market Defiant at this time. It is merely a proof-of-concept prototype for aerodynamic research."

DEFIANT MEETS THE PRESS

The Defiant first flew in June, 1978. Almost immediately it was shown to the aviation trade press and a mass of laudatory semi-technical information was available to every aviation enthusiast in the world. Full-color covers blossomed throughout the flying publications.

We had the opportunity to be one of the first handful of reporters to share the Defiant's cockpit with Burt Rutan. Actually, it was as a result of this first Defiant flight that Burt finally agreed to go ahead with the book project that you are now reading.

There were only 30 hours logged on the Defiant when Burt Rutan met us at Brackett Field, just east of Los Angeles. A group of interested spectators appeared out of the woodwork as he taxied in alone and shut down the engines (Fig. 10-6). Ground control had a list of questions as long as your arm. Here's an airplane that was *news* from the time it started on the drawing boards.

Before flight, Burt went over many of the design concepts that were outlined earlier in his *Canard Pusher*. Burt emphasized that the new "push-me, pull-me" design was nothing more than a "P.O.C." (proof of concept) test bed. Since this first exposure to

Fig. 10-3. Left side of the Defiant as it has been instrumented. The aircraft is fully IFR equipped, has redundant electrical and nav com systems.

Fig. 10-4. Right side of the Defiant's cockpit. Notice novel throttles at the bottom of the picture with the pointer on the throttle indicating the appropriate engine. Dual mixture controls are vernier knobs just in front of the throttles. The co-pilot's side stick is at the far right.

homebuilt enthusiasts, there have been many rumors of FAA certification program and eventual production. To date, neither has happened and the Defiant's future is still a great big question mark with the eventual answer somewhere in Burt Rutan's mind.

At one time, Rutan had decided to go one step farther with the Defiant concept and fight the FAA papermill on his own. He told his newsletter readers:

"We have decided to proceed with a type-certification program on a light twin based on the Defiant prototype. It's cabin design is completely different from the prototype. It has a clam-shell door with roomy seating and baggage for five adults. Specifics on its configuration, performance, etc., will not be released until it is in flight test in late 1980. Certification is anticiapted in late '81 or early '82. Meanwhile, we continue to gather operational data on the Defiant prototype, N78RA. Last week it flew nonstop from Mojave to Wichita (1265 miles) at 17,500 feet, at 175 knots, burning only 5.4 gallons per hour per engine, landing with 1 ½ hours fuel onboard. It is being evaluated in the IFR environment, including approaches at minimums, a zero-zero takeoff (acutal IMC), and light ice on four occasions. It has seen 270 hours of very vigorous testing of all types, including aerobatics and spin-attempts, and has yet to cut a flight short due to a precaution of any kind. Due to the fact that we are doing this program with very few people and very low overhead, we will not have time to respond to questions on the Defiant for the next two years. We plan to be very low key on release of any information or promotion of any type. The newsletter will report any releasable progress as we proceed."

In 1977-78, Rutan designed and built the Defiant in 7½ months with 1 ½ to 2 people working on it. He estimates between 1800 and 1900 man hours on the prototype. On a production line, he feels that the Defiant could be built for 350 man hours—about the same as a Cessna 150 or the Beech Duchess in production. Rutan estimates that he has $150,000 invested in the project.

Since the project began there have been a couple of "almost starts" toward FAA Certification and a production line. "Pug" Piper initally wanted to finance the development, but Burt decided to do it on his own. After the Defiant first flew "Pug" Piper wanted to set up a very large production, as Burt describes it, "with a lot of retired Beech and Piper people and 'improve' the aircraft, making it closer to the Baron or the Cessna Skymaster in terms of systems. I felt that was the wrong way to go. You can't or you'll wind up like the Lear Fan 2100 which may put composite construction back for years. The Lear needs a low parts count, not the 1400 present number of parts."

Then Burt met a multi-millionaire who offered to finance the project himself on a partnership basis, but he wanted the majority interest in the project for his money and Burt didn't think that was fair. "He was also a pilot and wanted to 'influence' the design," explained Burt; "things like retracting the gear. We didn't go that way."

On another "deal," the proposal was a joint venture with an established composite company in Southern California. This pro-

Fig. 10-5. Patricia Storch sits on the wing of the Defiant in front of the RAF at Mojave.

Fig. 10-6. The Defiant coasts to a stop on the ramp at Mojave after a flight demonstration.

posal had adequate financing and proper facilities, but the merger of talents didn't work out either. "We worked in 180° opposed positions," explained Burt. "We work in a direct way with not many people to make a few sheets of paper. They wanted many people to produce fewer sheets of paper."

There's really not much difference between the Defiant we flew on July 15, 1978, and the ship as it sits today tied down in the desert sun and sand so that the RAF hangar can be used for more pressing work (Fig. 10-7). A full set of nav/com equipment has been added, the rear seats and baggage compartment made habitable, and an external stop installed to eliminate gymnastics on entry and in deplaning.

However, the airframe and engines remain the same. No need has developed for the non-productive parts on conventional twins. There are still no tailcone, flaps, oil coolers or controllable props (Fig. 10-8). There are no engine nacelles and the main gear remains rigid. The wetted area (total aircraft surface) is only 50 percent of that of the Grumman Cougar.

As we discovered on our initial flight with Burt, the Defiant does not have the fatiguing out-of-sync propeller beat of most twins. There is a high damping effect of the fixed-pitch Kevlar wooden props.

In flight (Fig. 10-9), the main wing and winglets are far aft of your normal vision. It takes a look far around your shoulder to even assure that the second wing is following you. Aside from the canard in your field of view, flying the Defiant for me was much like riding the nose of a Northrop T-38 jet trainer where what little wing you have is way back there somewhere.

Back in mid-1978, there weren't all that many VariEzes in the air and the whole canard concept was completely new. It was easy to expect that the flight characterisitcs might be as different as the appearance of the airplane. They weren't.

When Burt demonstrated stalls out of steep climbing turns, there was a certain pucker factor as you waited for this weird bird to fall out of the air. It didn't take long to find out that the Defiant wouldn't stall nor would it spin.

With its clean design, it was a problem to slow down in the traffic pattern. A dive brake would help.

Single-engine procedures were completely straightforward and the number one safety item in the design. A high percentage of conventional twin-engine accidents are caused by slow-speed, engine-out operations, either in checkout or under actual conditions, when the off-center thrust of the "good" engine pulls the airplane up and over into a wild gyration that is frequently fatal. In single-engine operations with the Defiant, the main problem is ascertaining which engine is out. The front prop will stop only when the Defiant is slowed to 70 mph.

Just about the only thing that isn't goof-proof in the Defiant is the retractable nose gear. If you forget the gear on the VariEze or Long-Ez, about all that is hurt is your pride, a few coats of paint and maybe a little fiberglass ground off the bottom of the nose. With the Defiant, you'll also prang the front prop and ding the front-mounted "Rhino" rudder. Thus, Rutan has a carefully prepared check list, and he's used it from the beginning.

Aside from the problem in getting slowed down on final approach, landing with the tri-gear is a piece of cake. The canard's up

Fig. 10-7. The Defiant awaits better weather at Santa Fe, New Mexico. The author was ferrying the Decathlon in the background just before Christmas when weather toward the West Coast became impossible.

Fig. 10-8. Detail of the nose gear and "Rhino" rudder of the Defiant. The state-of-the-art push-pull is tied down outside the RAF buildings at Mojave while other work is done on the single-engine canards.

elevator is not strong on the ground and the nose wheel gets to the pavement very shortly after the mains. On that first flight, we shot landings at Chino and Brackett, much to the delight of tower operators and airport hangers-on. Even at this early stage, flying the Defiant was something to remember.

After this first flight, there was a personal feeling that we'd just jumped ahead into a new dimension of safe, simplified efficient flight. The Defiant, or one of its kinfolk, should be the new format for tomorrow's twin. We had that sort of a gut feeling after our first exposure to the Defiant—we still have that same feeling today.

FORECAST FOR THE FUTURE

The best forecast that Rutan has as this book is completed is that he will eventually form a new company to develop both the Defiant and the Predator (Rutan's 70' turbo-prop agricultural design concept).

"In the meantime, I want to take some time and really enjoy the Defiant," says Burt. "I want to take a couple of long trips, maybe to Europe with the Long-EZ's tank in the back seat. I'd like to join up with a couple of twins and go to Hawaii, then come back alone and maybe go as far non-stop as Dallas. I want to see if I have any 'young blood' left in my veins." [And Burt Rutan is only 37]

When not out on a brief demonstration, the Defiant is Burt's personal transportation. True, the RAF owns a Grumman Tiger, but if the trip calls for multi-engine safety, the Defiant is ready and willing to go.

With 350 flight hours at press time, the Defiant flies regularly. Whenever Burt has a special visitor, like Graham G. Murphy, a consulting engineer and airline captain from Papakura, New Zealand, in for a visit, the Defiant comes out of the hangar, the engines propped and the VIP visitor gets a first-hand view at what can very well be tomorrow's "twin."

"For some reason, that airplane has just been a cream puff," explained Burt with pride. "Really, about all we've done to the design has been to extend the rudder chord by 15 percent. The aircraft has required no repairs. Just change the oil and go (Fig. 10-10)."

The Defiant is thought by many well-qualified observers to be the most important single breakthrough in general aviation since the Beech Bonanza was first introduced in 1947. Between flights, it sits in Rutan's hangar at Mojave near Edwards AFB. This site is fitting since it dates back to the early state-of-the-art flights of the U. S. Army Force's Fokker tri-motor, "Question Mark" where the concept of air-to-air refueling was refined.

Not far away at Fox Field, Billy Barnes is rebuilding NR613K, a Model R Travel Air, sister ship of the famed "Mystery Ship" that was flown by his mother, "Pancho" Barnes to a women's speed record before WW II.

So, add the name of Rutan's Defiant to the Question Mark and the Mystery Ship. The main difference here is that the Defiant is not just a page of aviation history; it is surely destined for great things in the future.

Fig. 10-9. Defiant leads the way in a three-plane formation with the VariEze, center, and the Long-EZ on the outside of the grouping.

Fig. 10-10. Burt Rutan takes time off from his other projects to check the Defiant. As time permits, he just wants to "change the oil and go."

When we tried again to pin him down as to what his plans were for the Defiant, Burt's only comment was: "You'll just have to wait and see." But there was an aggressive gleam in his eye, much like the riverboat gambler who says, "I'll stay with these," and lays down a royal flush.

Appendix A
Specifications And
Performance Of Rutan Aircraft

	VariViggen	VariEze		Long-EZ	Quickie	Defiant
		N7EZ	N4EZ			
Engine	150 HP Lycoming	62HP VW	100 HP Continental	108 HP Lycoming	18 HP Onan	(2) 160 HP Lycoming
Wing Area (square feet)	123	59	66.6	94.1	50	127.3
Span	19'	21'	21.2'	26.3'	16'8"	29'2"
Empty Weight	1020	399	585	750	240	1585
Gross Weight	1700	880	1050	1325	480	2900
Fuel Capacity (gallons)	NA	14	28	52	8	90
Cabin Length	NA	95	100	100	64	NA
Cabin Width	NA	21.5	22.2	24.2	22	NA
Takeoff Distance	850	660	650	550	660	NA
Landing Distance	300	NA	1000	450	835	NA
Climb (fpm)	1200	1680	1900	1750	425	1750
Max Cruise Speed	165	175	196	185	127	188
Stall Speed	45	76	68	57	53	64
Ceiling	NA	NA	25,300	26,900	12,300	NA

Appendix B
Cost Comparison Of
Homebuilt Aircraft Engines

ENGINE	1 TIME AVAILABLE TILL NEXT OVERHAUL (REGULAR USE - 300 HR/YR)	2 OUTRIGHT COST FOR ENGINE	3 VALUE OF ENGINE AFTER RUNOUT	4 COST TO OWNER (2 - 3)	5 ENGINE COST PER HOUR OF FLYING
A65/A75 OSMOH	1200 hr	$1300	$400	$900	$0.75
A65/A75 half-runout	600 hr	$ 800	$ 400	$ 400	$0.66
C85/C90 OSMCH	1200 hr	$1900	$ 700	$1200	$1.00
C85/C90 half-runout	600 hr	$1300	$ 700	$ 600	$1.00
New 0-200	1800 hr	$4700	$1600	$3100	$1.72
0-200 OSMOH	1600 hr	$3000	$1200	$1800	$1.13
0-200 half-runout	800 hr	$2300	$1200	$1100	$1.38
Top of line VW conversion	800 hr?	$2600	$ 800	$1800	$2.25
"Low cost" VW conversion	100 to 600 hr?	$1300	$ 200	$1100	$2.20

Appendix C
Comparison Of "Quickie"
With Other Popular Light Aircraft

Aircraft	H.P.	Cruise @ 75%	m.p.g. @ 75%	Range @ 75%	Empty Wt.	Gross Wt.	Useful Load	Stall	R/C	Service Ceiling
Cessna 150	100	123	21.6	435	1046	1600	554	48	670	14,000
Bellanca Citabria	115	116	14.5	522	1097	1650	553	45	725	12,000
Piper Tomahawk	112	125	17.3	555	1064	1670	606	53	700	N.A.
Dehavilland DH60 Moth (1925)	60	80	17.3	345	770	1350	580	41	430	N.A.
Sorrell.Guppy	18	65	33	170	330	N.A.	N.A.	33	350	N.A.
Quickie	18	121	75	550	240	480	240	53	425	12,300
Piper J-3	65	75	19	220	690	1340	650	41	350	11,200
Whing Ding II	14	40	35	20	123	310	187	26	N.A.	4,000
Volksplane	40	75	30	240	440	750	310	46	400	N.A.
Pober Pixie	60	85	24	290	543	900	357	30	500	N.A.
Birdman TL-1	14	50	25	100	130	350	220	20	300	N.A.
Rand KR-1	40	150	48	420	355	600	245	45	600	12,000
Aerosport Quail	40	115	23	230	534	792	258	48	850	12,000
Teenie Two	40	110	40	310	590	280	50	800	15,000
Pazmarny PL-4	50	98	28	340	578	850	272	48	650	13.000

Notes: 1. All speeds and ranges are in Statute Miles
2. Except for the Dehavilland DH60, all data for this table was taken from information supplied by the respective designer/owner.

Index